FORAGING MARYLAND, VIRGINIA, AND WASHINGTON, DC

HELP US KEEP THIS GUIDE UP TO DATE

Every effort has been made by the author and editors to make this guide as accurate and useful as possible. However, many things can change after a guide is published—regulations change, techniques evolve, facilities come under new management, etc.

We appreciate hearing from you concerning your experiences with this guide and how you feel it could be improved and kept up to date. While we may not be able to respond to all comments and suggestions, we'll take them to heart and we'll also make certain to share them with the author. Please send your comments and suggestions to the following e-mail:

editorial@GlobePequot.com.

Thanks for your input!

FORAGING MARYLAND, VIRGINIA, AND WASHINGTON, DC

Finding, Identifying, and Preparing Edible Wild Foods

Christopher Nyerges

ESSEX, CONNECTICUT

FALCONGUIDES®

An imprint of Globe Pequot, the trade division of
The Rowman & Littlefield Publishing Group, Inc.
4501 Forbes Blvd., Ste. 200
Lanham, MD 20706
www.rowman.com

Falcon and FalconGuides are registered trademarks and Make Adventure Your Story is a
trademark of The Rowman & Littlefield Publishing Group, Inc.

Distributed by NATIONAL BOOK NETWORK

British Library Cataloguing in Publication Information available

Library of Congress Cataloging-in-Publication Data

Names: Nyerges, Christopher, author.
Title: Foraging Maryland, Virginia, and Washington, DC : finding,
   identifying, and preparing edible wild foods / Christopher Nyerges.
Description: Guilford, Connecticut : FalconGuides, [2022] | Includes index.
   | Identifiers: LCCN 2022009239 (print) | LCCN 2022009240 (ebook) | ISBN
   9781493058808 (paperback) | ISBN 9781493058815 (epub)
Subjects: LCSH: Wild plants, Edible--Middle Atlantic
   States--Identification.
Classification: LCC QK98.5.U6 N93855 2022 (print) | LCC QK98.5.U6 (ebook)
   | DDC 581.6/3209752--dc23/eng/20220329
LC record available at https://lccn.loc.gov/2022009239
LC ebook record available at https://lccn.loc.gov/2022009240

CONTENTS

ACKNOWLEDGMENTS

My primary ethnobotanical mentor was Dr. Leonid Enari, whom I first met in 1975.

Dr. Enari's unique background in botany and chemistry made him ideally suited as a primary source of information. He earned two higher degrees in his home country in both botany and chemistry (the equivalent of a PhD) in his twenties before emigrating to the United States from Estonia, where he experienced some of the results of Nazi occupation. He would tell his students that he pursued these fields because he desired to help people. "With the knowledge of botany and chemistry," he once told our class, "no one need ever go hungry."

When he first moved to the United States, he settled in Portland, Oregon, and taught at Lewis and Clark College and the University of Portland, where he wrote *Plants of the Pacific Northwest*. He eventually moved to Southern California, where he taught and researched for the remainder of his life.

Dr. Enari acted as my teacher, mentor, and friend, and he always encouraged me on to further research as well as teaching and writing. He assisted me with my first book, *Guide to Wild Foods*, and also consulted on various botanical writing projects.

I felt a great loss when he passed away in 2006 at age eighty-nine.

To Dr. Enari I dedicate this book, my eighth in the Falcon Guides Foraging series.

I also had many other mentors, teachers, and supporters along the way. These include (but are not limited to) Dr. Luis Wheeler (botanist), Richard Barmakian (nutritionist), Dorothy Poole (Gabrielino "chaparral granny"), Richard E. White (founder of WTI, who taught me how to teach and how to think; it was through WTI that I began teaching), John Watkins (a Mensan who "knew everything"), and Mr. Muir (my botany teacher at John Muir High School). These individuals all imparted some important aspect to me, and they have all been my mentors to varying degrees. I thank them for their influence. Euell Gibbons also had a strong influence on my early studies of wild food, mostly through his books; I met him only once.

Of course, there have been many others who taught me bits and pieces along the way, and I feel gratitude for everyone whose love of the multifaceted art of ethnobotany has touched me in some way. Some of these friends, associates, and strong supporters have included Pascal Baudar, Mia Wasilevich, Peter Gail, Gary Gonzales, Dude McLean, Alan Halcon, Paul Campbell, Rick and Karen Adams, Barbara Kolander, Jim Robertson, Timothy Snider, Dr. Wakemann, Dr. Muir, and many others.

I give special thanks to my beloved Helen for her support of this time-consuming project and because this book was her idea! While visiting my brother in Virginia, Helen thought that cataloging the wild foods of the greater DC area would make a great book, and our publisher agreed.

I wish to give special thanks to Malcolm McNeil, a professional naturalist who teaches in Virginia, for his active assistance with the development, the text, and the photos for this book.

I also thank the other naturalists-botanists who assisted with this project, such as Vickie Shufer, Tim McWelch, Rabiah Nur, Gina Richmond, and some others who chose to remain unnamed.

PHOTO CREDIT

Yes, I took many of the photos in this book, but I couldn't do it all myself. Rick Adams deserves special thanks. Other folks who contributed photos include my wife Helen, Zoya Akulova, Algie Au, Alex Chan, John Doyen, Barbara Eisenstein, Gary Gonzales, William J. Hartman, Barbara Kolander, Louis-M. Landry, Jeff Martin, Malcolm McNeil, Tom Nyerges, Jean Pawek, Jim Robertson, Vickie Shufer, Dr. Amadej Trnkoczy, Lily Jane Tsong, and Mark Vorderbruggen.

FOREWORD

My name is Malcolm McNeil. I am a naturalist/historian who lives in Fairfax County, Virginia, and works at Huntley Meadows Park, a Fairfax County Park Authority wildlife sanctuary. I describe my work as that of being an interpreter. I want to present factual information to you in an easy to understand way so you may improve yourself and feel fulfilled. This work is, hopefully, the first of many among many subjects I will be able to contribute to.

My relationship with Christopher and Helen Nyerges began several years ago on a lucky phone call that I was fortunate enough to have him answer. He came to be a guest speaker at one of my classes at the park, and I was blown away by his inherent kindness and in-depth knowledge. When we spoke afterward, I knew that I had found a kindred spirit in education, someone who wanted others to learn about the world around them. I am so proud to be able to contribute to this book and look forward to our continued relationship.

In this book, our goal is to provide a reliable source of information for anyone trying to learn about the forageable and useful plants that grow in the Virginia, Maryland, and DC areas. The different regions, climates, and ecosystems of these places will be addressed in order to provide a way to more easily identify plants of interest. This book will

Helen W. Nyerges, Malcolm McNeil, and Christopher Nyerges
KYLIE STARCK

not have specific sites, parks, or addresses to go to and harvest plants, but rather it will show how to identify plants around you as well as provide generalized areas and climates that certain plants may survive in.

Christopher and I want you to be able to look at the world around you with a more critical eye. I agreed to work on this project because I saw it as a way to continue providing people with a way to learn this valuable skill. Gaining the ability to connect more closely with the natural world through knowledge is powerful. Knowing about the natural world will help you learn about the human world as well. Once you learn about some of the history of plants, you recognize that humans have long intertwined their history with plants.

My background is in history, but alongside that, I hold a deep appreciation for the natural world. Through this education, I learned much about the Native Americans who lived in Virginia, Maryland, the DC area, and all over the United States. Learning about them was difficult, as much of their culture was wiped out by the colonists moving into the land. However, learning about their cultures taught me an important lesson. Much of their way of life was shaped by the nature around them. Their clothing, their tools, their ceremonies, and, of course, what they ate all shaped who they were. The land in Virginia, Maryland, and the DC area has changed over the course of many generations, but there is still life that has been here for many years that can be learned about.

The people who have been here all along, the Native Americans, are the ones who have provided so much knowledge to us. The Powhatans, Mattaponi, Pamunkey, Piscataway, Lenape, Rappahannock, Nottoway, and many more were all tribes that lived in this area. These individual tribes spoke unique languages and had different customs and cultures, but universally, they held a reverence for the earth around them. They learned what plant would soothe and which would hurt, which could be eaten after being boiled or which needed to be harvested when green. Their knowledge and ability to harness the world around them were incredible.

The Native Americans of the eastern woodland coast had a culture that was deeply entwined with plants and animals. Many of their homes were made with bent saplings and roofed by tree bark or reed mats. Their food was entirely that which they grew, hunted, or gathered for their tribe. Deer, turkey, rabbit, and fish were some examples of the hunted food.

The lessons passed on were done so orally. Oral history is extremely important when you do not have as many ways as today to record your information. These lessons were important to pass from one generation to the next. When people from faraway lands began arriving, there were generations upon generations of lessons passed down. These people were deeply familiar with the land around them. I like to think of this book, despite being written, as a way for us to "tell you" all these lessons.

Their ability to gather wild food from the land around them was not their only triumph. They also practiced agriculture, growing crops such as squash, corn, beans, and sunflower. Fields would sometimes look overgrown or neglected, but to a trained eye, all the plants were "cooperating" to a degree. The corn and sunflowers would serve as stalks for the beans to grow on, and the squash would spread along the ground and prevent weed growth. Only a people who had paid close attention to the land around them would be able to learn that lesson.

The colonists, enslaved people, and immigrants who came here to build up our country also brought with them many new plants. Sometimes labeled as weeds or invasives, plants from foreign areas can still be found, identified, and used to our own benefit. Learning who brought the plants here can help us understand how to use these plants again, while also encouraging us to experiment with new ways to use the plants.

I have found that studying the nature of this area taught me much about the peoples, just as studying the peoples taught me about the nature. Learning about edible wild plants, herbs for medicines, or fibers for tools teaches one how to look upon nature as a comfort rather than an unknown. A tool that you are unfamiliar with might seem scary, while one that you have learned becomes invaluable.

Hopefully, this book will be able to help you identify ways to harness some of the knowledge learned by the peoples who lived here before us. This helps preserve the nature as well as the way of life and culture of those before us. We cannot forget that they are still here as well. They continue to help educators like me provide a way to pass on their information to anyone who is willing to learn.

Another way to ensure that lessons continue to be passed on is keeping our natural areas protected and enduring. Harvesting wild foods can be harmful if done thoughtlessly, without care of whether or not there is enough left behind to reproduce. This book will be giving you different ways of harvesting the plants, making sure that you will not have to kill the plant to gain something from it. Then, that plant will be able to heal and give to another person like you. This giving, this responsibility, is important to make sure we can continue to live sustainably with our natural world, rather than taking to excess.

Our hope with this book is that anyone who reads it can feel a deeper connection with the world around them. In times of strife or danger, people have a tendency to fall back on the "old ways." Find some comfort in these "old ways" by learning what plants could enhance your daily living or save you in an emergency. It is not always life or death—it could be as straightforward as identifying a simple plant to add to a salad. The most important lesson is to keep an open mind and always be willing to learn. Thank you for coming along on this journey with us.

—Malcolm McNeil

INTRODUCTION

Types of Environments

Though this book is organized by Botanical Families, you will want to take note of the environments where each plant lives.

The areas that we use to categorize these plants include Ocean Side, Riparian, Marsh/Swamp, Lowlands/Valley, Higher Elevations, and Urban Development/Disturbed Soils. These environments are found all over Virginia, Maryland, and Washington, DC.

Ocean Side environments include areas that are close to saltwater expanses. The Atlantic Ocean along Maryland and Virginia's east coast is the most prevalent feature. These east coast areas are affected by differing soil quality and water availability.

Riparian areas are those near or affected by rivers. Rivers crisscross DC, Maryland, and Virginia, which leads us to the prevalent bodies of land affected by inland water. Generations ago, these waterways were highways for the people of the region, serving as trade routes and communication paths for many. These areas will have lots of aquatic plants.

Marsh/Swamp, along with wetlands and bogs, are areas that have incredible diversity in their plant life. These unique environments have shaped much of the way of life in the overall area. They are areas of changing water levels and many different aquatic plants, as well as being humid and wet overall.

Lowlands/Valley are prevalent as well, probably thanks to the high density of waterways in Maryland and Virginia. These areas are fertile and rich in biodiversity, full of a fantastic variety of plants. The Shenandoah Valley is a well-known example of a valley in Virginia.

Higher Elevations, found in the western parts of Virginia and Maryland, are home to strong trees and other sustainable plants. Usually having lower temperatures and higher winds, these areas are home to the hardiest of plants that will be discussed.

Urban Development/Disturbed Soils areas will include a lot of the plants found in the Washington DC area, and Maryland and Virginia also have their own metropolises. There is a surprising amount of biodiversity in these areas, including many plants that are written off as weeds or "useless." These include common plants you see on your everyday walk in town or weeds cropping up in your garden plot that may prove to be useful as food, drink, or medicine.

Scope of This Book

Foraging Maryland, Virginia, and Washington, DC intends to cover plants that can be used primarily for food and are common here. We are not attempting to cover every single edible plant that could possibly be used for food or those that are very marginal as food. Our focus is on those wild foods that are widespread, easily recognizable and identifiable, and are sufficient to create meals. Many of the wild edibles that are too localized or only provide a marginal source of food are not included. Plants that are endangered or have the possibility of being endangered have not been included. In general, plants that are too easily confused with something poisonous have also been omitted. Mushrooms are not included.

The content of this book is intended to be useful for hikers and backpackers, as well as for anyone in urban and rural areas where so many of these plants still grow. Our goal is to provide a book that details the plants that a person attempting to live off the land would actually be eating.

If you embark on the study of ethnobotany, and start working closely with a mentor/teacher, your learning will expand way beyond the pages of this book, and that is how it should be.

Organization

The plants in this book are organized according to the system used by botanists.

Many books on plants organize the plants by flower color or environmental niche, both of which have their adherents and their pitfalls. However, this book categorizes the plants according to their families, which gives you a broader perspective on many more plants than can be reasonably put into one book. As you will see, many of the genera (and some families) are entirely safe to use as food. Though there is no shortcut to learning about the identity and uses of plants, understanding the families will impart a far greater insight into the scope of "wild foods."

We'll start with the "lower" plants, then the gymnosperms (the cone-bearing plants), then the flowering plants, in alphabetical order by their Latin family names.

COLLECTING AND HARVESTING WILD FOODS

Since more and more people are desiring to learn how to "live off the land" and use wild plants for food and medicine, please practice sustainable collecting and harvesting methods.

The art (and science) of foraging is a sacred art, passed down from generation to generation, from the beginning of time. It is a fundamental skill, letting the fruits of the land feed you, as you, in turn, do everything you can to maintain the health and integrity of the land. People of the past instinctively knew that their survival was intimately tied to the health of the land. You could not overharvest, overgraze, and exploit without severe consequences. Remember: we and the land are One. It is an unfortunate point of modern days that this simple fact is being forgotten.

Always make sure it is both legal and safe for you to harvest the wild foods. Legality can usually be determined simply by asking a few questions or making a phone call. In some cases, when we're dealing with public lands, the issue of legality may be a bit more difficult to ascertain.

You also want to be safe, making sure there are no agricultural or commercial toxins near and around the plants you intend to harvest. Again, it pays, in the long run, to carefully observe the surroundings and to ask a few questions.

Foraging has gotten more popular in the last few years for some very good reasons. People want to reconnect with their roots and discover how our ancestors lived off the land. We're also discovering that wild foods not only taste good, but they are generally far more nutritious than just about everything from the modern supermarket.

So, when you forage, be sure to set a good example both in how you forage and in what you forage. Don't uproot plants if you don't need to. Don't overharvest from any area. And keep in mind that the landscape is often overrun with invasive nonnative plants, the so-called "weeds," mostly from Europe. There is usually a nearly unlimited number of these plants, and you are usually welcome to collect these plants.

There has been some backlash recently by those who feel foragers are ruining the landscape. I would agree that in some very narrow instances, this has been the case—typically, with mushroom collectors and, perhaps, with wildcrafters of herbs for the resale market.

I implore you to develop a relationship with the plants and learn how to harvest safely and sustainably. It is not that hard to be a conscientious forager, and in some cases, you might actually see an increase in the number, size, and health of wild plants by your very careful thinning, pruning, and collecting. If you consider how the human race moved from forager to farmer, it was precisely through this careful and intimate relationship with the plants—benefiting both parties—that early farming methods developed.

It is unfortunate that modern farming—far more than foraging—negatively affects vast swaths of native vegetation all over the globe, something we regard as a necessary aspect of food production. (That's another story, of course. However, we strongly encourage you to learn more about how food is produced these days. Some good references in this regard are *Everything I Want to Do Is Illegal* by Joel Salatin and *In Defense of Food* by Michael Pollan. Both books give you a good picture of modern food production, the many problems therein, and some practical solutions—one of which is to eat wild foods!)

Unless it is the root that you are using for food, you should never need to uproot a plant, especially if it is only the leaves that you intend to eat.

I have documented in my *Extreme Simplicity* book how I was able to extend the life of many annual weeds by carefully pinching back the leaves that I wanted to eat and then allowing the plant to grow back before picking again. Even when I believe that someone else will pull up the plant later or plow the area, I still do not uproot the plants on general principle. If I leave the plant rooted, the root system is good for the soil, and the plant continues to manufacture oxygen. Various insects and birds might eat the bugs on the plant or its seeds. Let life continue.

When you are harvesting greens, snippers can be used, but usually, nothing is needed but your fingernails and maybe a sharp knife. Cut what you need, don't deplete an area, and move on.

Harvesting seeds is done when the plant is at the end of its annual cycle, but there is still no reason to uproot the plant. When I harvest curly dock or lamb's-quarter seed, I carefully try to get as much into my bag as possible. I know that some seed is being scattered, and that's a good thing for next season. I also know that a few seeds are still on the stalk, and that's a good thing for the birds in the area.

I nearly always harvest in an area of abundance. If there are very few specimens there, my usual course of action is to simply leave them alone.

You will note when you read this text that I advise foragers to leave the wild onions in the ground and to eat the greens. In cases of abundance, your thinning the roots will help to stimulate more growth, and that is a good thing, akin to the passive agricultural practices of Native Americans.

In general, foraging doesn't require many tools. You will need bags, plastic, cloth, paper—whatever is appropriate for the food item. In some cases, you harvest with buckets or tubs. Usually, no other tools are needed, though I generally carry a clipper for any cutting and a knife or two. I rarely need a trowel, though it comes in handy with some harvesting.

The more you forage, the more you'll realize that your best tool is your memory. You'll learn to recognize where the mushrooms grow, where the berry vines are, and the fields that will be full of chickweed next spring. And the more you know, the less you'll need to carry.

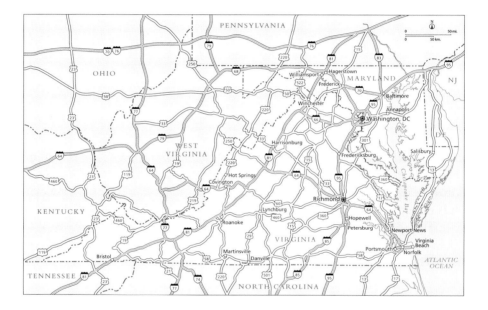

HOW MUCH WILD FOOD IS OUT THERE, ANYWAY?

Plants Everywhere, But Not All Can Be Eaten

In his book *Participating in Nature*, Thomas Elpel created a unique chart, based on years of observation and analysis, to give a perspective on the sheer numbers of edible, medicinal, and poisonous plants. Elpel is also the author of *Botany in a Day*.

First, almost every plant with known ethnobotanical uses can be used medicinally; even some otherwise toxic plants can be used medicinally if you know the right doses and proper application. So, yes, medicine is everywhere. But nearly two-thirds of these plants are neither poisonous nor used for food for various reasons.

The extremely poisonous plants that will outright kill you are rare. And since there are so few of these deadly plants, it is not all that difficult to learn to identify them. In our area and surrounding states, for example, there is poison hemlock that can be easily confused for something edible. Others that could cause death are various mushrooms and certain commonly planted ornamentals. It is not uncommon to hear about mushroom sickness and even death.

Though there are only a few plants that are deadly poisonous, many more would make you very sick but would not normally kill you. Still, all the poisonous and toxic plants combined are just a very small percentage of all the known ethnobotanicals.

Edible, Medicinal, and Poisonous Plants

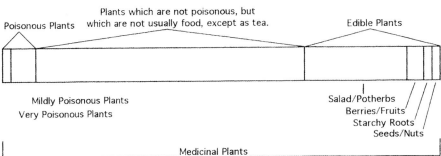

Reprinted with permission of Tom Elpel

Edible plants comprise maybe a quarter of the known edible, medicinal, and poisonous plants. Of the plants that we normally think of as "food plants," the overwhelming majority primarily provide us with greens. That is, throughout most of the year, the majority of the food that you'll obtain from the wild consists of greens: food to make salads and stir-fries and add to soups and vegetable dishes. These are plants that will not by themselves create a filling and balanced meal but will add vitamins and minerals to your dried beans, MREs (meals ready to eat), freeze-dried camping food, and other foods. In general, greens are not high sources of protein, fats, or carbohydrates.

Berries and fruits comprise another category of wild foods. A small percent of the wild foods you find will be berries or fruits, and timing is everything. Unlike greens, which you can usually find year round, fruits and berries are typically available only seasonally, so if you want some during other times of the year, you'll need to dry them or make jams or preserves. This includes blackberries, elderberries, mulberries, and many others. They provide sugar and flavor, but like greens, you would not make a meal entirely from fruits and berries.

Then, an even smaller category of wild foods than the berries consists of starchy roots, such as cattails and Jerusalem artichokes. These are great for energy, though they may not be available year round. Therefore, these foods have traditionally been dried, and even powdered, and stored for use later in the year.

Another small category of wild foods consists of seeds and nuts. This includes grass seeds, pine nuts, and acorns, among many others. It is in this small category where you obtain the carbohydrates, oils, and sometimes proteins that constitute the "staff of life." Though these are not available year round, some have a longer harvest time than others. Some may have a harvest period of as short as two weeks. Many grass seeds simply fall to the ground and are eaten by animals. Fortunately, most of these can be harvested in season and stored for later use.

ARE WILD FOODS NUTRITIOUS?

It is a common misconception that "wild foods" are neither nutritious nor tasty. Both of these points are erroneous, as anyone who has actually taken the time to identify and use wild foods can testify. I've had many new students who had been convinced about the nutritional value of wild foods but had assumed that the plants nevertheless tasted bad. Of course, a bad cook can make even the best foods unpalatable. And if you pick wild foods and don't clean them, don't use just the tender sections, and don't prepare them properly, you'll almost certainly turn people off to wild foods.

Wild foods are not only nutritious but can be as flavorful as any foods in the finest restaurants.

For your edification, here is a chart extracted from the USDA's *Composition of Foods* to give you an idea of the nutritional content of the common wild foods.

Nutritional Composition of Wild Foods (per 100 grams, unless otherwise indicated)

Blanks denote no data available; dashes denote lack of data for a constituent believed to be present in measurable amounts. Only a select number of plants for which we had data are represented. Primary source: Composition of Foods, US Department of Agriculture.

Plant	Calories	Protein (g)	Fat (g)	Calcium (mg)	Phosphorus (mg)	Iron (mg)	Sodium (mg)	Potassium (mg)	Vitamin A (IU)	Thiamine (mg)	Riboflavin (mg)	Niacin (mg)	Vit. C (mg)	Part
Amaranth	36	3.5	0.5	267	67	3.9	—	411	6,100	0.08	0.16	1.4	80	Leaf, raw
Carob		4.5	0.7	352	81	2.9	35	827	14		0.4	1.89	0.2	Pods
Cattail		8%	2%											Rhizomes
Chia		20.2%		631	860	7.72	16	407	54	0.62	0.17	8.8	1.6	Seed
CHICORY TRIBE														
Chicory	20	1.8	0.3	86	40	0.9	—	420	4,000	0.06	0.1	0.5	22	Leaf, raw
Dandelion	45	2.7	0.7	187	66	3.1	76	397	14,000	0.19	0.26	—	35	Leaf, raw
Sow thistle	20	2.4	0.3	93	35	3.1	—	—	2,185	0.7	0.12	0.4	5	Leaf, raw
Chickweed														
Dock	28	2.1	0.3	66	41	1.6	5	338	12,900	0.09	0.22	0.5	119	Leaf, raw
Fennel	28	2.8	0.4	100	51	2.7	—	397	3,500	—	—	—	31	Leaf, raw
Filaree	—	2.5	—	—	—	—	—	—	7,000	—	—	—	—	Leaf
Grass										300–500 IU	2,000 to 2,800 IU		300 to 700 mg	Leaf, raw
Lamb's quarter	43	4.2	0.8	309	72	1.2	43	452	11,600	0.16	0.44	1.2	80	Leaf, raw
Mallow	37	4.4	0.6	249	69	12.7	—	—	2,190	0.13	0.2	1.0	35	Leaf
Milkweed	—	0.8	0.5	—	—	—	—	—	—	—	—	—	—	Leaf
Miner's lettuce						10% RDA			22% RDA				33% RDA	Leaf
MUSTARD FAMILY														
Mustard	31	3	0.5	183	50	3	32	377	7,000	0.12	0.22	0.8	97	Leaf
Shepherd's purse	33	4.2	0.5	208	86	4.8	—	394	1,554	0.08	0.17	0.4	36	Leaf
Watercress	19	2.2	0.3	120	60	0.2	41	330	3,191		0.12	0.2	43	Leaf
Nasturtium														
Nettle	65	5.5	0.7	481	71	1.64	4	334	4,300	—	0.16	0.38	76	Leaf
New Zealand spinach	19	2.2	0.3	58	46	2.6	159	795	4,300	0.04	0.17	0.6	30	Leaf, raw
Oak (acorn flour)	65% carbohydrates	6%	18%	43	103	1.21	0	712	51	0.1	0.1	2.3	0	Nut

Plant	Calories	Protein (g)	Fat (g)	Calcium (mg)	Phosphorus (mg)	Iron (mg)	Sodium (mg)	Potassium (mg)	Vitamin A (IU)	Thiamine (mg)	Riboflavin (mg)	Niacin (mg)	Vit. C (mg)	Part
ONION FAMILY														
Chives	28	1.8	0.3	69	44	1.7	—	250	5,800	0.08	0.13	0.5	56	Leaf, raw
Garlic	137	6.2	0.2	29	202	1.5	19	529	—	0.25	0.08	0.5	15	Clove, raw
Onion	36	1.5	0.2	51	39	1	5	231	2,000	0.05	0.05	0.4	32	Young leaf, raw
Passion fruit [per pound]				31	151	3.8	66	831	1,650				71	Fruit
Pinyon	635	12	60.5		604	5.2				1.28				Nut
Prickly pear	42	0.5	0.1	20	28	0.3	2	166	60	0.01	0.03	0.4	22	Fruit, raw
Prickly Pear	16	1.2	trace	163	17	0.7	22	319	415	0.01	0.04	0.5	13	Pad
Purslane	21	30	1.7	0.4	103	39	3.5	—	—	2,500	0.03	0.1	0.5	Leaf & stem, raw
Rose	162	1.6		169	61	1.06	4	429	4,345		0.16	1.3	426	Fruit, raw
SEAWEED														
Dulse	—	—	3.2	296	267	—	2,085	8,060	—	—	—	—	—	Leaf
Irish moss	—	—	1.8	885	157	8.9	2,892	2,844	—	—	—	—	—	Leaf
Kelp	—	—	1.1	1,093	240	—	3,007	5,273	—	—	—	—	—	Leaf

Ferns

There are 13 families of ferns. According to Dr. Leonid Enari, the young, uncurling, growing tips of all ferns are edible once any hairs are removed and when they are cooked. They taste a bit nutty. Most are also edible raw. Many have a long history of being steamed and served with butter or cheese or mixed into various vegetable dishes.

There are many ferns that you will encounter in our area, besides what we have presented here. The bracken has a long history of use as food.

BRACKEN FAMILY (DENNSTAEDTIACEAE)

Among the ferns, the Bracken family contains about 11 genera and about 170 species.

Bracken leaf RICK ADAMS

BRACKEN
Pteridium aquilinum

Use: Young uncurling shoots used for food
Range: Throughout the area, mostly in the shady areas of the hills and canyons
Similarity to toxic species: See Cautions
Best time: Spring
Status: Somewhat common in the correct terrain
Tools needed: Clippers

Properties

Bracken can apparently be found worldwide and throughout the United States. In this area, bracken can be found throughout the states in pastures, hillsides, wooded areas, and even in full sun. You'll find it most commonly on the northern, shady side of hillsides or shady hillsides where water seeps and where little sun gets through the canopy of whatever larger trees grow there.

The rhizomes are hairy and sprawling underground, sometimes branching. The petiole is black near the base, with dense brown hairs. The plants grow from one to four feet tall, and the overall appearance of each frond is roughly triangular; each is twice-pinnately divided.

Uses

The young shoots are the edible portion, and they have the appearance of the head of a fiddle, which is where the common name "fiddlehead" comes from. The young shoots will uncurl and grow into the full fern fronds. The young tips are picked when young and can be eaten raw or cooked. I like to toss a few in salads when the fiddleheads are in season; they impart a nutty flavor.

More commonly, these are boiled or steamed and served with butter or cheese. They are easy to recognize and have gained a resurgence of popularity as more people are rediscovering wild foods. Bracken is also a good vegetable to add to soups and stews and mixed dishes.

Just carefully pinch off the tender unfolding top, and you can gently rub off the hair. Use as a nibble or cook. Do not eat the fully opened ferns.

Cautions

Researchers have identified a substance called ptaquiloside in bracken fern, a known carcinogen. So, is it safe to eat? It has been a food staple of Native Americans for centuries, if not millennia, and the Japanese also enjoy bracken and consider it one of the delicacies of spring. Although actual scientific data is

Bracken fiddlehead BARBARA KOLANDER

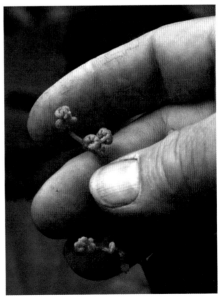

A small fiddlehead RICK ADAMS

inconclusive, there is a higher rate of intestinal cancer among Native Americans and the Japanese, and this could be linked to the use of bracken fern. Livestock has been known to be mildly poisoned by eating quantities of the raw bracken ferns. Cooking is known to remove some of the toxins, though not necessarily the ptaquiloside.

Despite this, there are many who are not so concerned about such inconclusive studies. For example, Steven Brill, in his *Identifying and Harvesting Edible and Medicinal Plants*, states, "I wouldn't be afraid of eating reasonable quantities of wild [bracken] fiddleheads during their short season." Another forager, Green Deane, says, "I am willing to risk a few fiddleheads with butter once or twice a spring, which is about as often as I can collect enough in this warm place."

The final choice is up to you. For perspective, we regularly hear things far worse than the above about coffee, high-fructose corn syrup, sugar, and french fries, yet people seem to have no problem purchasing and eating these substances. That doesn't make them good for you, but eating some in moderation is not likely to be the sole cause of cancer or other illness.

My mentor, Dr. Leonid Enari (holder of doctorate degrees in chemistry and botany), regarded the entire category of the fiddlehead stage of ferns as safe for food. He stated that he knew of none that were toxic if you heed the following precautions: cook all fiddleheads that you intend to eat, since some may be a bit toxic raw. He advised cleaning fiddleheads of hairs, if any, before cooking. Dr. Enari also advised to not eat any mature fern fronds. Though many may be safe when mature, they are not as palatable as the young fiddlehead. Thus, you need to get to know that individual fern before you eat its mature fronds. Otherwise, eat only the fiddleheads, clean them of hairs, and cook them before eating.

Others who I spoke to in preparing this book told me that they hold a much more conservative viewpoint regarding the edibility of *all* fern fiddleheads and would only eat those species that have been identified as edible.

Some Other Fern Groups

The Spleenwort family (Aspleniaceae) includes about 10 genera worldwide, with 650 species. This includes the ostrich fern (*Matteuccia struthiopteris*), a distinctive fern that can grow from two to six feet tall. These are native to the Eastern United States, as well as parts of Europe. These are popular with chefs in the United States, Japan, and elsewhere. The young fiddleheads are covered with brown scales that should be rinsed or scraped off before cooking.

Gymnosperms

This is a class of plants whose seeds are formed in cones (as with pine trees), or on stalks (as with the genus Ephedra). The members of this group include the cycadophytes, conifers, ginkgo tree, and the Ephedras.

PINE FAMILY (PINACEAE)

A view of pine needles and their cones LILY JANE TSONG

PINE
Pinus spp.

There are seven species of Pinus in our area.

Use: Needles for tea and spice; nuts for food

Range: Various species are found in the mountains and throughout the area. Often planted in urban areas.

Similarity to toxic species: None

Best time: Nuts in the fall; needles can be collected anytime

Status: Common in certain localities; often planted

Tools needed: Clippers for needles

Properties

Pines are fairly widespread trees, with some preferring burned-over areas. The Pine family is said to supply about half the world's lumber needs. The family consists of 10 genera and 193 species. There are 94 species of *Pinus* in the Northern Hemisphere, and at least seven are known to grow in this area.

Uses

Though there are a few *potential* foods with the pines, it is primarily the seeds that will provide you with food that is both substantial and palatable.

The cones mature and open in the fall. As the scales open sufficiently, the seeds drop to the ground, where they can be collected if you're there at the right time and beat the animals to them. The seeds may drop over a two-week to a month period. One of the best methods to harvest is to lay sheets under the trees to catch the seeds so they're not lost in the grass. The seeds are then shelled and eaten as a snack, added to soups, or mashed and added to biscuits or pancakes.

FORAGER NOTE: Some of the very long needles of certain pines are excellent for coiled baskets.

I have taken the not-fully-mature cones and put them into the fire, carefully watching them so they don't burn. The idea is to open the scales and then get the seeds. Yes, you can get seeds this way, and you'll lose some too. I do not recommend this method.

The tender needles can also be collected and brewed into a tea. Put the needles in a covered container, and boil at a low temperature for a few minutes. This tea is rich in vitamin C and very aromatic and tasty—that is, if you enjoy the flavor of a Christmas tree, which is what you'll smell like after drinking it. It's very good.

Yes, we have all heard of eating the cambium layer of pine trees. I regard this as a "survival food," meaning it could be worth all the work involved if you're actually near starving, but it also depends on the species of pine whose cambium you are collecting, the location, and time of year. Some people have had positive experiences with this food source.

Pine nuts in the shell and shelled

Magnoliids

This category was formerly considered part of the dicot group of the broader category of Angiosperms; some botanists still regard it as a sub-category of eudicots. Though eudicots make up 75 percent of known species of angiosperms, the magnoliids make up only 2 percent. Magnoliids have two cotyledons (like eudicots) and flower parts in multiples of three, like monocots. The number of furrows or pores on pollen is one, like monocots.

LAUREL FAMILY (LAURACEAE)

Mitten-shaped leaves of the young sassafras plant HELEN W. NYERGES

SASSAFRAS
Sassafras albidum

Use: Root is used for a beverage; leaves used in gumbo
Range: Somewhat widespread
Similarity to toxic species: None
Best time: Any time
Status: Common
Tools needed: Shovel or clippers

Properties

This native is very common as a medium-sized tree. It is easy to recognize from its irregularly lobed leaves, which could be oval or one- or two-lobed. We often think they look like mittens! The smooth leaves lack teeth and are alternately arranged. The bark of the older trees is reddish brown, irregularly furrowed, and thick. The young twigs and smaller branches have a smooth green bark. The entire plant is very aromatic.

Though very widespread—you can even find it on urban streets—I have most often seen it in the woods, especially on the edges of clearings and alongside fencerows and roadside ditches. You can find it in sunny and partially sunny locations.

Uses

Sassafras is a tree or shrub with a fairly unique appearance. Its leaves are quite distinctive. I remember seeing the plant for the first time along a road that went through the back of my grandfather's farm. The bushes were almost weedy in appearance, growing thickly where they had been cut back by someone trying to clear the road. The leaves were unmistakable!

I tried to pull a root, and only a portion came out. I smelled it and thought it was the sweetest, most appealing aroma I'd ever smelled. It was unique and reminiscent of both root beer and licorice. Later, I learned why: root beer was so called because the original root beer—a far cry from today's artificially flavored, white-sugared drinks called "root beer"—was actually brewed from roots, including sassafras, ginger, sarsaparilla, and, sometimes, ginseng.

I took my little root home and brewed the entire thing in water, and I drank the hot beverage without sweetening. I found it deliciously satisfying. I later learned that most people brew just the bark of the root, not the whole root, though it really doesn't seem to matter.

The bark of a mature sassafras tree
MALCOLM MCNEIL

A view of the distinctive sassafras leaves
MALCOLM MCNEIL

Gumbo

Leaves of sassafras have long been dried and powdered, and once the stems are removed, they are used as a thickening agent for soups and stews.

Cautions

Sassafras and sarsaparilla both contain safrole, a compound banned by the FDA due to its carcinogenic effects. Safrole was found to contribute to liver cancer in rats when given in high doses, and thus, it and sassafras or sarsaparilla-containing products were banned. Sassafras's carcinogens are about a tenth that of alcohol, and in our research, we're unaware of anyone who ever contracted liver cancer from periodic consumption of sassafras root tea or from the use of gumbo from the leaves.

A sassafras can become a large tree MALCOLM MCNEIL

Eudicots

This category was formerly referred to as dicots in the broader category of Angiosperms. The sprouts begin with two cotyledons, and the flower parts generally occur in fours and fives.

All families in this category are arranged alphabetically by their Latin name.

MUSKROOT FAMILY (ADOXACEAE)

This family has five genera and about 200 species worldwide.

David Martinez examines the ripening elderberry fruit.

ELDERBERRY
Sambucus spp.

There are 20 species of *Sambucus* worldwide. There are two species of *Sambucus* in our area, *S. canadensis* and *S. racemosa,* which has red fruits.

Use: Flowers for tea and food; berries for "raisins," jam, jelly, juice

Range: Throughout the area in the mountainous areas, urban fringe, and generally most environments

Similarity to toxic species: See Cautions

Best time: Early spring for flowers; early summer for fruit

Status: Common

Tools needed: Clippers for flowers; clippers and good, sturdy bucket for berries

Properties

Elder trees can be found throughout the United States and throughout our area. They can be found in drier regions, as well as along streams, and in the higher mountain regions. They are generally small trees, with oppositely arranged, pinnately divided leaves with a terminal leaflet. Each leaflet has a fine serration along its edge.

The plant is often inconspicuous for much of the year but is
very obvious when it blossoms in
its many yellowish-white flower
clusters in the spring. By early
summer, the fruits develop in
clusters, which are often droop-
ing from the weight.

Uses

Remember: "Black and blue,
is good for you; red as a brick,
will get you sick!" (This is an old
Boy Scout saying)

The blue berries, rich in
vitamin A, with fair amounts of
potassium and calcium, can be
eaten raw or can be mashed and
blended with applesauce for a
unique dessert, especially if you
are using wild apples. The berries can also be used for making
wines, jellies, jams, and pies.

Elderberry fruits

Though some of the Indig-
enous people ate the red berries
when cooked, there are people
today who get sick from the red
ones. I do not advise that you
eat red elderberries at all; however, if you decide to try them
anyway, cook them well and

FORAGER NOTE: If you don't want
your fruit to get all smashed and
crushed, don't collect in a bag. Collect
in a basket or bucket, and don't pack
too many into the bucket.

sample only a little bit at first to see how your body reacts.

Wild-food researcher Pascal Baudar likes to dry and powder the blue fruit
and sprinkle it over ice cream. The whole flower cluster can be gathered, dipped
in batter, and fried, producing a wholesome pancake. Try dipping the flower
clusters in a batter of the sweet yellow cattail pollen (see Cattail) and frying it
like pancakes. It's delicious!

Another method to use the flowers is to remove them from the clusters
and the little stems, and then mix with flour in a proportion of fifty-fifty for
baking pastries, breads, biscuits, and more. The flowers also make a traditional

Appalachian tea that is said to be useful for colds, fevers, and headaches.

The long, straight stems of elder have a soft pith and have historically been hollowed out and used for such things as pipe stems, blowguns, flutes, and straws for stoking a fire.

Cautions

Be sure to cook the fruit before eating it and avoid the red berries entirely. While not everyone will get sick from eating the dark purple or black berries raw, they can cause severe nausea in some people. Therefore, cook all fruit before using it for drinks or other dishes.

Do not consume the leaves, as this will result in sickness.

Collected elderberry, ready to have the stems removed, and then made into juice and jam

A view of the elderberry's flower cluster, fruit cluster, and leaf HELEN W. NYERGES

Elderberry Sauce

This simple sauce goes well with any game, such as duck, but feel free to try it with chicken too!

1 pound elderberries (freeze the clusters first; crush lightly with your hands, and the berries will fall easily)

1 large sweet onion or 7–8 scallions

²/₃ cup red wine vinegar

³/₄ cup sugar or honey

1 teaspoon grated ginger

A couple of cloves

¹/₂ teaspoon of salt, or to taste

Place the berries in a pot and squeeze them with your hand to release the juice. Place all the other ingredients in the pot and bring to a boil for 10 minutes, then strain the liquid through a sieve.

Return the liquid to the pot, bring to a simmer, and reduce until you have achieved the right consistency (like a commercial steak sauce). You can prepare the sauce in advance and keep it in the fridge for many days.

—RECIPE FROM PASCAL BAUDAR

HIGH-BUSH CRANBERRY
Viburnum spp.

There are about 200 species worldwide of *Viburnum*; in our area, we find *V. opulus*, *V. edule*, and others.

Use: The fruits are eaten
Range: Found throughout the area, including hedgerows on farms and even in the urban areas. The native species is found in moist, cool woods and swamps.
Similarity to toxic species: No
Best time: Flowers in May to July; fruit following in late summer
Status: Common
Tools needed: A collecting basket

A view of the fruit and leaf of V. opulus ZOYA AKULOVA   The fruit and leaf of V. opulus ZOYA AKULOVA

Properties

These are deciduous, semi-erect shrubs that can get up to 10 feet tall but are usually around four to five feet tall. The opposite leaves are petiolate, palmately veined, and shallowly three-lobed, appearing somewhat like a currant leaf. Sometimes you'll see some unlobed leaves. Each leaf is sharply toothed, usually three to 10 centimeters long and wide, with a pair of glandular projections near the junction with the petiole. The leaves turn conspicuously red in the fall, and by winter, the leaves fall off. The white flowers are formed in compound umbels. Each corolla is widely bell shaped, whitish, with five lobes. The fruit is a one-seeded drupe, and it matures to a bright red or orange.

Uses

When you encounter this semi-erect shrub, you will be inclined to taste a fruit if it is in season. You'll chew on it and spit out the seed, and maybe you'll like it, maybe you won't. I've heard people describe the flavor as musky, but let's just say it has a unique flavor. Cooking the fruits to make a jam or jelly will mellow the flavor, and most people will enjoy it at that point. I nibble on the raw fruit sometimes, but I prefer making a jam or juice from it. I do this by mashing the ripe fruits, pouring through a sieve, and then gently cooking the juice and sweetening it if I just want a juice. We've also cooked it down and used it as a pie filling.

You could also simply mix the high-bush cranberry fruit with other wild or domestic fruits for juices, jams, pie fillings, and so on.

Fruits from the often cultivated *Viburnum opulus* tend to be more bitter. The native *Viburnum edule* tends to be sweeter. Fruits that are left on the bush into the winter will be a bit mellower and are tasty raw.

AMARANTH FAMILY (AMARANTHACEAE)

The Amaranth family has 75 genera and 900 species worldwide. Of the members of the *Amaranthus* genus in this area, *A. retroflexus* seems to be the most common of about 10 species.

Rick Adams examines a tall erect amaranth in seed.

AMARANTH
Amaranthus spp.

Use: Seeds for soup or pastries and bread products; leaves can be eaten raw or cooked

Range: Amaranth is widespread. Common in the disturbed soils of farms, gardens, fields, open areas, and urban lots.

Similarity to toxic species: Some ornamentals resemble amaranth. Some toxic plants superficially resemble amaranth, such as the nightshades (e.g., *Solanum nigrum*, whose raw leaves could get you sick). Individual jimsonweed leaves

FORAGER NOTE: Amaranths are a diverse group. Some have an erect stalk, some are highly branched, and some are prostrate.

(*Datura* spp.) have been confused for amaranth leaves. Generally, once the amaranth begins to flower and go to seed, this confusion is diminished.

Best time: Spring for the leaves; fall for the seeds

Status: Common

Tools needed: Tight-weave bag for collecting the seeds

Properties

Though there are many species of *Amaranthus*, *A. retroflexus* is most common in the area.

Amaranth is an annual. The ones with erect stalks can grow up to three feet and taller, depending on the species. Some are more branched and are lower to the ground. When young, the root of one of the common varieties, *A. retroflexus*, is red,

One of the erect amaranths

and the bottoms of the young leaves are purple. The leaves of *A. retroflexus* are oval shaped, alternate, and glossy. Other *Amaranthus* leaves can be ovate to linear.

The plant produces flowers, but they are not conspicuous. They are formed in spike-like clusters, and numerous shiny black seeds develop when the plant matures in late summer. The plant is common and widespread in urban areas, fields, farms, backyards, and roadsides.

Uses

Amaranth is a versatile plant with edible parts available throughout its growing season.

The young leaves and tender stems of late winter and spring can be eaten raw in salads, but because there is often a hint of bitterness, they are best mixed with other

A view of the young red root of Amaranthus retroflexus

A. retroflexus. Note the red root. RICK ADAMS

greens. Young and tender stems are boiled in many parts of the world and served with butter or cheese as a simple vegetable. Older leaves get bitter and should be boiled into a spinach-like dish or added to soups, stews, stir-fries, and so on. In Mexico, leaves are sometimes dried and made into flour, which is then added to tamales and other dishes.

Amaranth begins to produce seeds in late summer, and once the seeds are black, they can be harvested. The entire plant is generally already very withered and dried up by the time you're harvesting. The seeds are added to soups, bread batter, and pastry products.

The seeds and leaves are very nutritious; no doubt, this is part of the reason this plant was so revered in the old days. One hundred grams of the seed contains about 358 calories, 247 mg of calcium, 500 mg of phosphorus, and 52.5 mg of potassium. The seed offers a nearly complete balance of essential amino acids, including lysine and methionine.

The leaf is also very nutritious, being high in calcium and potassium. One hundred grams, about a half a cup, of amaranth leaf has 267–448 mg of calcium, 411–617 mg of potassium, 53–80 mg of vitamin C, 4,300 mcg (micrograms) of beta carotene, and 1,300 mcg of niacin. This volume of leaf contains about 35 calories.

Historical note: The seed and leaf of this plant played a key part in the diet in pre-colonial Mexico. The seeds would be mixed with honey or blood and shaped into images of their gods, and these images were then eaten as a "communion." Sound familiar? After the Spanish invaded Mexico, they made it illegal to grow the amaranth plant, with the justification that it was a part of "pagan rituals."

Young A. retroflexus leaves RICK ADAMS

SUMAC FAMILY (ANACARDIACEAE)

The Sumac family includes about 70 genera and 850 species worldwide.

Christopher observes a large sumac tree growing on the grounds of the National Museum of the American Indian. HELEN W. NYERGES

STAGHORN SUMAC
Rhus typhina

There are 850 species of Rhus worldwide, with four found in the greater DC area.

Use: The fruit clusters are made into a beverage

Range: Found throughout the area, including farms, mountainous areas, and even the urban areas

Similarity to toxic species: When not in flower and fruit, bears a superficial resemblance to poison sumac

Best time: Flowers May to July; fruit following in late summer into the fall

Status: Common

Tools needed: A collecting basket

Properties:

Staghorn sumac is native to eastern North America. It is primarily found in southeastern Canada, the northeastern and midwestern United States, and the Appalachian Mountains. It is commonly cultivated as an ornamental. We've seen it on the edges of farms, in the wild at Shenandoah Park, and even on the Mall at the National Museum of the American Indian.

The plant is easiest to recognize when the conspicuous fruit is present in the fall, which sits like a red cardinal in the tree. The shape of the fruit—which is a cluster of the seeds—has been described as torch-like and about eight inches in length, more or less. The individual seeds are red and covered in hair.

The leaves are pinnately divided, with anywhere from 11 to 31 leaflets. The leaves are arranged opposite each other. The leaves are dark green above; the underside of the leaf is pale green and velvety.

The margins of the leaflets are serrated. The whole tree is very conspicuous in the fall as the leaves turn yellow or red. The bark is smooth and covered in velvety hair.

Uses

When I lived on my grandfather's farm, the staghorn sumac bushes were somewhat common in the "back 40." The fruits were conspicuous, and there's really

The fruit cluster is often on the tree after all the leaves have dropped.

nothing else you would confuse them with. My brother and I had read in one of Euell Gibbons's books about making "lemonade" from the red fruit clusters, and we were eager to try them.

We first collected the red clusters and gently rinsed them to remove any dirt. Then, I simmered them in warm water, strained out the water, and let it chill in the refrigerator. The juice had a sugar content but is usually quite tart at this point. I sweetened it with honey.

There are other ways that lemonade has been made from the fruit clusters, such as simply soaking the clusters in cold water, straining, and then sweetening the water to taste.

A cluster of the staghorn sumac fruit

Where there are many of these bushes growing, you could have a good beverage source for at least part of the year.

For those of you who enjoy an occasional smoke but do not want the harmful effects of tobacco, you might follow one of the practices of local Native Americans, who smoked the leaves of staghorn sumac. The leaves could be smoked alone but would usually be mixed with other smokable herbs and smoked in a pipe.

Some beekeepers use dried sumac bobs as a source of fuel for their smokers.

All parts of the staghorn sumac, except the roots, can be used as both a dye and as a mordant. Because the plant is rich in tannins, leaves and twigs can be added to other dye baths to improve lightfastness. The leaves may be harvested in the summer when available, and the bark can be harvested year round.

Viewing the staghorn sumac bush along the road after the leaves have dropped

CARROT (OR PARSLEY) FAMILY (APIACEAE)

The Carrot family has about 300 genera worldwide, with about 3,000 species. Many are cultivated for food, spice, and medicine, but some are highly toxic. Never eat anything that looks carrot- or parsley-like if you haven't positively identified it.

A patch of wild carrots

WILD CARROT
Daucus carota

There are about 20 species of *Daucus* worldwide, and at least one is found in this area.

Use: The roots can be dug and used like farm-grown carrots, but they are tougher. Seeds are sometimes used as a spice.

Range: Not a native, but can be found throughout much of the DC, Maryland, and Virginia areas, often in disturbed or poor soils and along roadsides

Similarity to toxic species: See Cautions

Best time: For a more tender root, dig before the plant flowers

Status: Found sporadically throughout the area

Tools needed: Shovel

Properties

If you've ever grown carrots in your garden, you will recognize the wild variety in the wild, except it will typically be smaller. Smell the crushed leaf, and scrape a bit of the root and get a whiff of it. Does it smell like carrot? The root of the wild carrot is white, not orange, and if you scratch it, you'll get that very characteristic carrot aroma.

The leaves are pinnately dissected, with the leaflet segments linear to lanceolate—just like the leaves of a garden carrot. The flowers are formed in umbels of white flowers, with a tiny central flower that is purple. As the umbel matures, it closes up and has the appearance of a bird's nest.

The whitish root of the wild carrot ALGIE AU

Note the purple flower in the middle of the carrot inflorescence. ALGIE AU

Uses

The roots can be dug, washed, and eaten like garden carrots, though they are usually tough. This was one of the first wild foods that I learned to use. It was always a feeling of great mystery and pleasure to dig into a wild field and pull up one of these white roots. Was it a farm carrot gone feral? Probably. It still woke up some distant inner memory of "hunter-gatherer."

Sometimes, I simply peel the outer part of the taproot and discard the very tough inner core. I then wash the outer layer of the root, chop or slice it, and add it to raw or cooked foods.

The entire root might be tender if it's growing in moist and rich soil. Then, the whole root can be sliced thin and added to salads. It is probably best added to soups, stews, and various cooked dishes.

But most of the roots I've found are pretty tough, and only the outer layer can be eaten. I slice this thin and cook it. It adds a good flavor to soups, though it lacks the carotene of commercial carrots.

The seeds of the mature plant can be collected and used as a seasoning, to taste.

Cautions

The Carrot family contains some very good food and spices; however, make absolutely certain that you have a carrot and not poison hemlock, which are both members of this family. The wild carrot plant has that very distinctive carrot aroma and has fine hairs on the stalk. Poison hemlock does not have a distinctive carrot aroma, and its stem has distinctive purple blotches or markings.

A view of the parsley-like poison hemlock leaf. Note the red to purple blotches on the lower stalk.

When young, poison hemlock resembles Italian parsley and looks very much like wild carrot as it grows taller. If uncertain, here are some tips: Rub a leaf and smell the aroma. Does it smell like carrot? Then it probably is. Does it smell musky, like dust or old socks? It's probably poison hemlock. Look at the mature stalk. If you see purple blotches, you have poison hemlock. Look at the flower umbel. Is there a single purple flower in the middle? If

The purple blotches of the lower stalk of poison hemlock

so, you have a wild carrot. Remember, never eat any wild plant until you have positively identified it.

The cow parsnip plant in flower BOB SIVINSKI

COW PARSNIP
Heracleum maximum, formerly *H. lanatum*

There are about 80 species of *Heracleum* worldwide; at least *H. maximum* is found in this area.

Use: The tender parts of the stalk are edible

Range: Prefers lowlands and moist watershed areas. Common near the mountains.

Similarity to toxic species: See Cautions

Best time: Spring

Status: Somewhat common

Tools needed: Knife

Properties

This is a robust plant, an obvious member of the Carrot or Parsley family because of the white flowers that are clustered in umbels. The flowers are composed of five sepals, five petals, five stamens, and one pistil.

The plant can grow up to 10 feet tall, though four to six feet seems to be the norm. It produces leaves with three large, coarsely toothed lobes, very much palmate, almost like maple leaves. The plant has a stout, hollow stalk. There is also a carrot-like taproot.

The plant is most commonly found in mountain meadows and moist areas. It's a typically conspicuous plant, which you cannot help but notice.

Uses

The tender stalk can be peeled and eaten raw, though it is really better when you cook it. The young stalks are best, and the flavor is often compared to celery.

The dried and powdered leaves have been used as a seasoning for other foods, generally as you'd use salt. You can experiment and see if this appeals to you. The leaves have also been dried and burned, and the ash is used as a seasoning.

Native Americans had many uses for the cow parsnip plant besides eating the young peeled stems. They would make the plant into a poultice to treat sores and bruises. The hollow stems were also used as drinking straws, as well as flutes.

Cautions

Although this is one of the easiest members of the Carrot family to identify, the family does contain some deadly members, so do not eat any part of this plant until you've made a positive identification.

Also, the young stalk is typically peeled before eating because the surface of the stalk causes a dermatitis reaction in some people.

Closeup of the cow parsnip stem ZOYA AKULOVA

The fennel plant with its stout stalks and fern-like leaves

FENNEL
Foeniculum vulgare

Use: Leaf and stalk raw or cooked; seed for tea or seasoning
Range: Common as an invasive species. Found along the coast and common locally in urban lots and fields.
Similarity to toxic species: Fennel has needlelike leaves and smells like licorice, so you really shouldn't confuse it with anything toxic. However, this family contains some poisonous and toxic members, so be certain you're picking fennel before eating it.
Best time: Spring is best for the young shoots; collect the seeds in summer or fall
Status: Common in certain localities
Tools needed: None

Properties

Fennel is the only species of the *Foeniculum* genus.

Fennel is a perennial from Europe that is common in some areas. Generally, it is considered invasive.

The plant begins to produce its leaves in the spring. The finely dissected leaves give the plant a ferny appearance. The base of each leaf clasps the stalk with a flared base, similar to the base of a celery stalk. The unmistakable characteristic is the strong licorice aroma of the crushed leaf.

In winter and early spring, the plants begin to appear. They first establish a ferny, bushy two-to-three-inch broad base. By spring and early summer, the flower stalks rise to a height of six feet (higher in ideal conditions). The entire plant has a slightly bluish-green cast due to a thin, waxy coating on the stalks and leaves. The yellow flowers form in large, distinctive umbels.

Uses

Young fennel leaves and peeled stalks are great to eat as a trail snack when you're thirsty and hungry. When the plant first sprouts in the spring, you can eat the entire tender, succulent base of the plant, somewhat like you'd eat celery. As it

Nathaniel Schleimer examines young fennel plants in his yard.

sends up its stalk but before the plant has flowered, the stalk is still tender and can be easily cut into segments. These tender segments are hollow and round in the cross section and can be used like celery for dipping, or cooked like asparagus and served with cheese and butter.

Later, as the plant grows taller, you can eat the tender leaves and stems chopped up in salads or added to soups and stews. It gets a bit fibrous as it matures but can be diced up and added to many dishes. It adds a sweet spiciness to the dishes in which it is used. If you don't care for licorice, you probably won't care for fennel.

As the plant matures, it sends up tall stalks.

The younger leaves are the best and sweetest; you can still eat the older leaves, diced fine, and sometimes, there is a slightly bitter or astringent under-flavor when eaten raw.

When the seeds mature, they can be made into a licorice-flavored tea. Just put five or six seeds into a cup and add hot water.

The seeds alone can be chewed as a breath freshener or used to season other dishes.

If you enjoy fennel, this is one of those ideal plants to grow in the lazy person's garden. It seems to take care of itself, does well in sun or shade, and continues to rise from its roots year after year.

DOGBANE FAMILY (APOCYNACEAE)

The Dogbane family has about 200 to 450 genera and from 3,000 to 5,000 species worldwide.

Milkweed stalk with flower buds

MILKWEED
Asclepias spp.

The genus *Asclepias* contains about 100 members, with about 11 found in this area.

Use: Young shoots, leaves, pods are boiled and eaten

Range: Widespread in both disturbed soils and some wilderness areas

Similarity to toxic species: Could bear a resemblance to some dogbanes

Best time: Collect the shoots early and the leaves before the plant flowers. The pods are available in early summer.

Status: Common

Tools needed: A bag, clippers

Properties

Milkweed is found throughout the United States. There are several species of milkweed across the country, and *A. syriaca* (common in the East) is almost certainly the one most commonly eaten.

This plant grows erect, from two to five feet tall, and is stout and whitish-green in color. It is usually found in patches. The stalk is fibrous and usually erect. Thick, milky white sap oozes out when the stalk or leaf is broken or cut. The ovate-to-oblong leaves are opposite or whorled, not toothed at the margins, and tapered at both ends, measuring up to five inches long.

The white, pink, or rose/purplish flowers are about half an inch in diameter, arranged in umbels. Each flower consists of a five-petaled corolla; each petal has an erect cowl and inwardly hooked horn. There are five sepals, five stamens, and two pistils with a superior ovary.

The blossoms develop into rough green pods (fruits), which are conspicuous and about three inches long. Their surface is covered with soft, spiny projectiles. When the pods mature, they split along one seam, revealing the neatly packed seeds. To each seed is attached a number of down-like silky fibers.

A patch of milkweed

The milkweed plant in the field

The mature milkweed fluff starting to blow in the wind HELEN W. NYERGES

Uses

The tender young sprouts, below six inches tall, can be eaten. When the plant is older, the leaves and flower clusters are also used for food. The immature seed pods, before they've developed the silky fibers inside (cut one open and inspect), are great in soup or with meat. All parts of the milkweed must be boiled in water before they are rendered palatable.

Originally, I followed Euell Gibbons's cautious warnings to always double and even triple boil milkweed before eating because it was said to be so bitter. However, just do your own taste test before eating. Sometimes, one boiling will do the job.

Cautions

According to Poison Control, "All parts of the plant contain toxic cardiac glycosides, which can cause nausea, diarrhea, weakness, and confusion in small amounts, and seizures, heart rhythm changes, respiratory paralysis, and even death in large amounts. Milkweed can also irritate the skin and eyes if touched." It sounds really bad, but we know of no cases of death ever occurring, and most human sicknesses were mild. Still, milkweeds should never be eaten uncooked. *A. speciosa* has resulted in livestock poisoning because they consume it raw in the field. However, many animals tend to avoid eating it.

Mature pods of milkweed with seed starting to fall out HELEN W. NYERGES

Dogbane Might Be Confused for Milkweed

Milkweed and dogbane are placed into the same botanical family (by most botanists), and so they have some obvious similarities. They both have white sap, and they can be found in the same environment. But there are some easy ways to tell them apart.

Sometimes the dogbane plant (above) is confused with milkweed.

COMPARING MILKWEED TO DOGBANE

	Milkweed (Asclepias syriaca)	Dogbane (Apocynum cannabinum)
Found where	Farmlands; fields; appears in spring, up to 7,000-foot level	Farmlands; fields; appears in spring, up to 7,000-foot level
Leaf	Broad elliptical leaf, covered in fine hairs. Arranged opposite each other; leaves clasp stalk.	Broad leaf, about as long as a milkweed leaf but narrower. No fine hairs. Arranged opposite; leaves clasp stalk.
Stems and stalk	Main stalk is hollow; white sap when cut. Plant has one erect stalk. Green stalk.	Main stalk is solid; white sap when cut. Upper part of plant branched. Green stalk; turning red when mature.
Flowers	Clusters of five-petalled pink to purple flowers	Clusters of little white flowers with five united petals
Fruits	Bulging seed pods, with warty surface. Splits open to reveal brown, flat seeds, each attached to silky fiber.	Pods are long and narrow, produced in pairs, each about three inches; drooping.
Sap	White sap when leaf or stalk is cut	White sap when leaf or stalk is cut
Fiber	Acceptable fiber for cordage; degrades in a few months	Quality fiber for cordage; quality persists for months to years

According to the USDA, dogbane is considered toxic to humans. Cymarin, a chemical found in the plant's roots, was used as a cardiac stimulant and was listed until 1952 in the medicinal text *United States Pharmacopoeia.* The milky sap contains cardiac glycosides (a chemical compound derived from a simple sugar and often of medicinal importance) that have physiologic actions similar to digitoxin.

PIPEVINE FAMILY (ARISTOLOCHIACEAE)

There are 10 genera of this family, with about 600 species worldwide.

The wild ginger plant in flower KEIR MORSE

WILD GINGER
Asarum canadense

There are 90 species of *Asarum* worldwide.

Use: Root used or leaves; used as one would use cultivated ginger, mostly as a seasoning or medicine.

Range: This plant prefers moist and shady areas. Found in all the Appalachian and Piedmont counties, as well as half of the coastal counties.

Similarity to toxic species: No

Best time: Spring to late fall

Status: Somewhat common

Tools needed: Trowel

Properties

The entire plant is no more than about 10 inches tall.

This is a perennial, whose leaves rise from a branching horizontal rhizome. The pair of leaves have long stems which rise directly from the root. The leaves are more or less heart shaped or arrowhead shaped and are sometimes mottled.

The leaf and leaf stalks are covered in fine hairs.

The plants are often somewhat conspicuous where they grow because of the pair of heart-shaped leaves. Sometimes they are solitary, and you might also find them in patches. At times, they are hidden by the leaves of the forest floor. These leaves are aromatic.

The flowers are solitary on each leafless stalk. The flowers, which appear in late spring, are brownish-purple with three sepals and no petals, 12 stamens, and one pistil.

A view of the wild ginger plant KEIR MORSE

Uses

This is used as you would use commercial ginger root, as a candy, seasoning, or medicinal plant.

If you've used ginger for cooking, you'll have no trouble working with this root.

The root is easily dug and can be used fresh or dried for later use. If you want to make a ginger candy, you slice the root into small slices and boil in water until tender. To make the candy, follow any candy recipe from a good cookbook.

You can also simply soak the root in any sweetened water and then use the water for cooking or for drinking as is.

Slice the slender roots into thin segments, and use in Chinese recipes, either fresh or dried. The dried root can also be ground fine and added to salad dressings or pickling spices.

If made into a candy, or infusion, this is said to be effective in soothing sore throats as well as for relieving stomach pains and intestinal disorders.

Yes, we admit this is a marginal food, though it often catches the eye because of the heart-shaped appearance of the leaves. Please don't over-harvest unless there's an abundance of this plant.

Cautions

Consume wild ginger sparingly or as an initiative novelty. The plant contains aristolochic acid, which can damage the kidneys. This acid is not readily water-soluble, so confections made from steeping the plant may be more benign.

SUNFLOWER FAMILY (ASTERACEAE)

Worldwide, the Sunflower family has about 1,500 genera and about 23,000 species!

Jepson divides this very large family into 14 groups. Most of the plants addressed here are in Group 7, described as having ligulate heads, five-lobed ligules (five teeth per petal), and, generally contain milky sap when broken. When I was studying botany in the 1970s, my teachers described this group as "the Chicory Tribe," a much more descriptive title than the unimaginative "Group 7."

According to Dr. Leonid Enari, the Chicory Tribe contains no poisonous members and is a worthy group for further edibility research. I have eaten many of the other members of this group not listed here, and generally, their bitterness requires extensive boiling and water changing to render edible and palatable.

GROUP 4

BURDOCK
Arctium spp.

Burdock growing by the side of the road HELEN W. NYERGES

There are 10 species of Arctium throughout Europe, and only two species have been recorded in this area.

Use: The root, stems, and leaves can be eaten, but the root is most commonly used

Range: Not a native but can be found throughout the area. Prefers old orchards, waste areas, and fields.

Similarity to toxic species: Resembles a rhubarb leaf

Best time: Best time to dig the root is in the first year's growth.

Status: Relatively common

Tools needed: Shovel

Properties

Wild burdock is found throughout this area and throughout most of the United States. The first time I saw wild burdock, I thought I was looking at a rhubarb plant, though the stalk was not red and celery-like, as with rhubarb.

The first-year plant produces a rosette of rhubarb-type leaves; in ideal soil, the second-year plant produces a stalk six to nine feet tall. Both of these species —*A. lappa* and *A. minus*—are similar, with *A. lappa* growing a bit taller.

The leaves are heart shaped (cordate) or broadly ovate. The leaves are conspicuously veined. The first-year leaves are large and up to two feet in length. In the second season, the plant sends up a flower stalk with similar but smaller leaves. The purple to white flowers, compressed in bur-like heads, bloom in July

The cut burdock root—also known as gobo—being prepared for cooking

and August. The seed containers are spiny-hooked burs that stick to socks and pants.

Burdock's root looks like an elongated carrot, except that it is white inside with a brownish-gray skin that is peeled away before eating. You sometimes find this root in the markets sold as "gobo."

A view of the burdock leaf

Uses

The first-year roots can be dug, washed, and eaten once peeled. They are usually simmered in water until tender and cooked with other vegetables. In Russia, the roots have been used as potato substitutes when potatoes aren't available. The texture and flavor of boiled burdock roots are unique, though it does resemble a potato. The roots can also be peeled and sliced into thin pieces and sautéed or cooked with vegetables. I also eat young tender roots diced into salad, and I find them very tasty.

Leaves can be eaten once boiled; in some cases, boiling twice is necessary, depending on your taste. Try to get them very young. Peeled leaf stems can be eaten raw or cooked. The erect flower stalks, collected before the flowers open, can be peeled of their bitter green skin and then dried or cooked, though these tend to be much more fibrous than the leaf stems.

An analysis of the root (100 grams or half a cup) shows 50 mg of calcium, 58 mg of phosphorus, and 180 mg of potassium. Tea of the roots is said to be useful in treating rheumatism.

Herbalists all over the world use burdock: the roots and seeds are a soothing demulcent, tonic, and alterative (restorative to normal health).

According to Linda Sheer, who grew up in rural Kentucky, burdock leaf was the best herbal treatment that her people used for rattlesnake bites. Two leaves are simmered in milk and given to the victim to drink. The burdock helps to counteract the effects of the venom. The body experiences both shock and calcium loss as a result of a rattlesnake bite. The lactose in the milk offsets the calcium loss and prevents or reduces shock. (I'd love to hear from chemists about how specifically burdock helps a rattlesnake victim recover.)

You can also take the large burdock leaves and wrap fish and game in them before roasting in the coals of a fire pit. Foods cooked this way are mildly seasoned by the leaves.

The flowering thistle plant RICK ADAMS

THISTLE
Cirsium spp.

Worldwide, there are about 200 species of the *Cirsium* genus. At least eight are found in this area.

Use: Edible stems, youngest leaves
Range: Found throughout the United States and fairly widely in our area
Similarity to toxic species: None.
Best time: Spring
Status: Widespread
Tools needed: Knife, clippers, bag

Properties
There are many species of thistle, all with very similar appearances. Thistles normally reaches four to five feet at maturity. They can be either perennial or biennial herbs.

Thistle leaves are alternate leaves, prickly or spiny, and generally toothed. They're about eight inches long at maturity. Thistle flowers are clustered in bristly heads. They are crimson, purple, pink, and occasionally white. The lower half of the flower heads are covered with spiny bracts, resembling thistle's cultivated relative, the artichoke.

Uses

When the plant is young and no flower stalk has emerged, the root can be dug, boiled, and eaten. These are starchy roots, mild-flavored, and as the plant flowers, the root becomes tough and fibrous. If you want to try eating the roots, you want to search in the spring in rich soil. If the soil and the timing are not right, the root is too tough to eat.

A young thistle rosette

My favorite part of the thistle is the stalk, cut before the flowers develop while the stalk is still tender and perhaps about three feet tall. Using a sharp knife, I first cut off the bristly leaves and then carefully scrape off the spiny outer layer on the stalk so I can handle it. Once the stalk is scraped of its outer fibrous layer, you can eat it raw—it is sweet and somewhat reminiscent of celery. We've served it with peanut butter for a tasty snack.

These cleaned stalks can be baked, boiled, sautéed, or added to other foods. They can be served with butter or cheese, as you might serve asparagus.

If you've ever seen a garden artichoke, then the relationship of artichoke to thistle might be obvious. Can you eat the flowering head of the thistle like you'd eat an artichoke? Well, it depends. First, you have to clip the thistle flower while it's still young and before it has actually flowered. You boil it, and then if you peel back all the scaly bracts, you'll find just a little bit of the tender heart there. It's very tasty and worth trying, but usually, there is really not that much heart to the wild thistle to make it worth your bother.

In general, I don't eat the leaves because I find the

The youngest thistle growth is the best to eat. RICK ADAMS

A thistle plant sending up its flowering stalk

prickliness irritating. I was once surprised when I was in the field with fellow forager Lanny Kaufer, who pulled out a small pair of scissors when we came across thistle. Within about 15 seconds, he trimmed the prickles off the perimeter of the leaf and ate it raw.

I prefer the very youngest leaves of spring, which would be tasty even in a salad if chopped fine.

On occasion, when the timing was right, I was able to collect many of the very young emerging cotyledons and add them to salads and cooked dishes. Since the flavor of the leaves is a bit sweet, you could collect young leaves and process them in a juicer that removes the pulp and have a tasty and healthy drink.

GROUP 5

CHAMOMILE
Matricaria chamomilla

There are seven species of *Matricaria* worldwide; only three are found in our area.
Use: The entire plant can be used for a pleasant tea

A view of the chamomile flowers VICKIE SHUFER

Range: Widespread, preferring hard soils

Similarity to toxic species: There are many members of the Sunflower family that are small and might be confused for Chamomille to the beginner. Remember the unique aroma of chamomille, and that this plant prefers hard-packed soil in the wild.

Best time: Spring, when the flowers are present

Status: Prefers hard-packed soils

Tools needed: Sharp knife or scissors, bag

Properties

Chamomile is easily cultivated in pots or in gardens. It readily goes feral and seems to prefer hard-packed soil, such as gravel driveways, the type of soil where you cannot easily stick a shovel.

This is an annual herb that can rise as much as eight inches, but it is usually less than that—typically three or four inches tall. The leaves are finely divided into short, narrow linear segments, which are

The individual leaf

alternately arranged and glabrous (not hairy). The flower heads are formed at the ends or tips of the branches and are cone shaped and small, about half an inch in length. The ray flowers ("petals") are white.

When you crush the leaves and particularly the flower head, you get a distinctive sweet aroma.

Uses

When I lived to the north at my grandfather's farm in Ohio, chamomile grew wild in the hard-packed soil between our garage and farmhouse. Each day, when I came home, I walked the short distance to the house, an unpaved area where chamomile thrived. Even when not in flower, the air was perfumed by my walking on the carpet of chamomile! It was a pleasant experience, even though I tried my best to not walk on these delicate plants.

Chamomile growing in a planter

Mostly, chamomile is used to make an aromatic beverage. It is normally used for its reputed calming effects, even though chamomile doesn't seem to have the more noticeable sedative effects of a tea such as passionflower. Many people commonly use it as a beverage simply because they enjoy its flavor.

The whole aboveground herb can be cut and infused to make a pleasant tea. Or, you can snip off just the flowers to make an even more flavorful beverage. Some people find the flavor of the entire herb a bit bitter, but this might vary from plant to plant. I generally use the entire herb, largely because it's time-consuming to pick each flower head.

You can dry this herb for later out-of-season use, or you can use it fresh when it's in season. I like it best fresh, but that's not possible year round. Drink plain or sweetened, as you wish.

The main constituents of chamomile flowers are polyphenolic compounds, and essential oil components from the flowers are terpenoids. Chamomile has long been regarded as an herb to cause relaxation, reduce anxiety, and treat insomnia. Chamomile tea has antioxidant, anti-inflammatory, and astringent properties. According to a Case Western Reserve study, chamomile has been proven to help reduce symptoms of the common cold, gastrointestinal conditions, and throat soreness and hoarseness. It's widely touted as a sleep aid and an effective home remedy to reduce anxiety. Yes, most of this is anecdotal evidence and not based on clinical studies. Still, an infusion of these flowers continues to enjoy great popularity.

Cautions

Since these low-growing plants are so close to the ground, be sure to wash them well before using. People allergic to ragweed might experience an allergic reaction when drinking chamomile tea.

Lindsay Champion wrote a 2020 blog on wellness for the *PureWow* website, asking if it's safe to drink chamomile tea while pregnant. She writes, "We polled several obstetricians, and the general consensus is that drinking chamomile tea is a personal decision you should make with your doctor. There is no hard-and-fast rule as to whether or not chamomile is definitely safe or definitely unsafe. Because there is so little research in regard to pregnant women and chamomile tea, it's best to err on the side of caution."

She points out that researchers aren't permitted to experiment on pregnant women, and so no one really knows. At least there appears to be no evidence that chamomile tea is harmful to a pregnant woman. So just like with any new food, try a little if you're so inclined, and monitor your body's reactions.

Chamomile tea

GROUP 7 (or 8, depending on which botanist's system you follow)

CHICORY
Cichorium intybus

A view of the overall chicory plant RICK ADAMS

There are about six species of *Cichorium* worldwide; only this one is found in our area.

Use: Root for beverage and food; greens raw or cooked

Range: Widespread throughout the United States; found especially in the disturbed soils of farms, fields, and gardens

Similarity to toxic species: None

Best time: Spring

Status: Common locally

Tools needed: Digging tool for roots

Properties

The chicory plant grows upright, typically three to five feet tall, with its prominent sky-blue flowers. Look carefully at the flower—each petal is divided into five teeth, typical of the Chicory Tribe of the Sunflower family. Each leaf will produce a bit of milky sap when cut. The older upper leaves on the stalk very characteristically clasp the stem at the base.

This is a perennial from Europe that is now widespread, mostly in fields, gardens, disturbed soils, and along roadsides.

Uses

This is another of those incredibly nutritious plants with multiple uses. The leaves can be added to salads, preferably the very young leaves. If you don't mind a bit of bitterness, the older leaves can be added to salads too. The leaves can be cooked like spinach and added to a variety of dishes, such as soups, stews, and egg dishes.

A view of the chicory leaf RICK ADAMS

Chicory roots are also used, either boiled and buttered or sliced and added to stews and soups. Roots in rich soil tend to be less woody and fibrous.

Chicory roots have long been used as a substitute for coffee or as a coffee extender. Dig and wash the roots, and then dry them, grind them, and then roast them until they are brown. Use as you would regular coffee grounds, alone or as a coffee extender. Incidentally, you can make this same coffee substitute/extender with the roots of dandelion and sow thistle.

Note: The entire Chicory Tribe of the Sunflower family contains no poisonous members, though many are bitter. These are generally tender-leafed plants with milky sap and "dandelion-like" flowers, each petal of which usually has five teeth at the tip.

Gary Gonzales shows a can of a coffee and chicory blend.

PRICKLY LETTUCE
Lactuca serriola, et al.

A young rosette of prickly lettuce, still tender enough for salad

There are about 100 species of *Lactuca* worldwide, with six recorded in our area. *Lactuca serriola*, a European native, is probably the most abundant and widespread.

Use: Young leaves, raw or cooked

Range: Found throughout the United States. Most commonly found in gardens, disturbed soils, along trails and edges of farms.

Similarity to toxic species: None

Best time: Early spring

Status: Widespread

Tools needed: Bag for collecting

FORAGER NOTE: One of the common names for this plant is "compass plant." When the plant is mature, the edges of the leaves tend to point to the sun as it moves across the sky. This is probably a mechanism to stop water loss. While this is by no means as accurate as using a compass, it could help you determine directions with a bit of figuring.

Properties

Prickly lettuce is a very common annual plant that you can find just about anywhere, hidden in plain view. Yes, you've seen it, but it likely blended into the landscape. It's mostly an "urban weed," though occasionally you'll find it in the "near wilderness" surrounding urban areas.

Prickly lettuce rises with its erect stalk to no more than three feet. The young leaves are lanceolate with generally rounded ends. They are tender, and if you tear a leaf, you'll see white sap. The plant grows upright with an erect stem, which develops soft spines as it gets older. As the plant matures, you'll note that there is a stiff line of hairs on the bottom midrib of the leaf. The leaf attachment is either sessile or clasping the stem, and the leaf shape can be quite variable, from a simple oblong-lanceolate leaf to one that is divided into pinnately lobed segments. Despite this, after you've seen a few prickly lettuce plants, you should readily recognize it.

The flowers are small and dandelion-like, pale yellow, with about a dozen ray flowers per head. As with dandelions, these mature into small seeds attached to a little cottony tuft.

Uses

Prickly lettuce sounds like something you'd really like in a salad, but in fact, you need to find the very youngest leaves, or they get too tough and bitter. Very young leaves (before the plant has sent up its flower stalk) are good added to your salads and sandwiches.

Stiff spines on the bottom midrib of the leaf. This is why you don't eat the mature prickly lettuce leaves!

The stalk of the maturing prickly lettuce plant

The leaves can also be collected and mixed into stir-fries or added to soups or any sort of stew to which you would add wild greens.

But let's not be fooled by the name "lettuce." Yes, it's botanically a relative of the cultivar you buy in the supermarket, but the leaves become significantly bitter as they age, and the rib on the underside of each older leaf develops stiff spines that make any similarity to lettuce very distant. This means you'll be using this plant raw only when it's very young, and when it's flowering and mature, you probably won't be using it at all.

Still, it's edible, and it grows everywhere. You should get to know this plant, and its relatives, and learn to recognize it early in the growing season.

I've used it when *very* young in sandwiches, salads, soups, stews, and egg dishes. I've even used the small root when I was experimenting with coffee substitutes. Since prickly lettuce is related to dandelion and sow thistle, I figured it would work well as a coffee substitute, and it does, but there's very little root to this plant.

SOW THISTLE
Sonchus oleraceus, et al.

The sow thistle plant in flower

There are about 55 species of *Sonchus* worldwide. Three are found in our area, all of which are from Europe and are edible.

Use: Mostly the leaves, raw or cooked; root can be cooked and eaten; flower buds pickled

Range: Found throughout the United States; most common in urban areas, gardens, and farms but can be found in most environments

Similarity to toxic species: None

Best time: Spring, though the older leaves of late summer are still useful

Status: Common

Tools needed: Trowel for digging

Properties

Though the common sow thistle (*S. oleraceus*) is most commonly used for food, the other two species found in this area look very similar and can be used the same way. When you see *S. asper*, the prickly sow thistle, you may conclude that it's too much work to use for a dish of cooked greens because it is covered with soft spines.

FORAGER NOTE: Sow thistle is one of our most common wild foods. It is so widespread that it can be found in nearly every environment, even in the cracks of urban sidewalks.

When most people see a flowering common sow thistle for the first time, they think it's a dandelion. Yes, it is related to the dandelion, and yes, the flowers are very similar.

Here is a simple distinction: all dandelion leaves rise directly from the taproot, forming a basal rosette. Sow thistle sends up a much taller stalk, up to five feet or so in ideal conditions but usually about three feet. The leaves are formed along this more or less erect and branching stalk. The leaves are paler and more tender than the dandelion leaves, and sow thistle leaves are not as jagged on the edges as dandelion. Though the individual dandelion and sow thistle flowers are very similar, dandelion only forms one flower per stalk, whereas sow thistle will form many flowers per stalk.

A bee collecting nectar from the sow thistle flower

Uses

Though sow thistle may not be quite as nutritious as dandelion, it's definitely tastier, and the leaves are more tender. You can include the leaves of sow thistle in salads, and even when the plant is old, there is only a hint of bitterness. The flavor and texture are very much like lettuce that you might grow in your garden.

The leaves and tender stems are also ideally added to soups and stews or simply cooked up by themselves and served like spinach greens. They are tasty alone, or you can try different seasonings (such as peppers, butter, and cheese) that you enjoy.

The root can be eaten or made into a coffee substitute, as is commonly done with two of its relatives, dandelion and chicory. To eat the roots, gather the young ones and boil till tender. Season as you wish and serve. The roots could also be washed and added to soups and stews.

The flowering sow thistle in an urban backyard

For a coffee substitute, gather and wash the roots, then dry thoroughly. Grind them into a coarse meal, roast to a light shade of brown, and then percolate into a caffeine-free beverage. Is it "good"? It's all a matter of personal preference.

RECIPE

Spring Awakening

For a dish that resembles asparagus, take just the tender sow thistle stems in the springtime (the leaves can be removed and added to other dishes). Boil or steam the stems until tender—it doesn't take long—and then lay on your plate like asparagus. Add some cheese or butter, and it will make a delicious dish, but one that you'll only enjoy in the spring—timing is everything.

DANDELION
Taraxacum officinale

Dandelion's bright yellow composite flower

There are about 60 species of *Taraxacum* worldwide, with at least two of these found in this area.

Use: Leaves raw or cooked; root cooked or processed into a beverage

Range: Common nationwide; prefers lawns and fields and disturbed soils

Similarity to toxic species: None

Best time: Spring for the greens; anytime for the roots

Status: Common

Tools needed: Trowel for the roots

Properties

Even people who say they don't know how to identify any plants can probably identify a dandelion in a field. The characteristic yellow composite flower sits atop the narrow stem, which rises directly from the taproot. There is one yellow flower per flower stalk. These mature into the round, puffy seed heads that children like to blow on and make a wish.

Dandelion's rosette of leaves

Dandelions grow in fields, lawns, vacant lots, and along trails. They tend to prefer disturbed soils, and they tend to prefer urban fields, though I have seen them in the wilderness.

The leaves are dark green, toothed on the margins, and each rises from the root. The name "dandelion" actually comes from the French *dent de leon*, meaning "tooth of the lion," for the jagged edges of the leaves.

Uses

My first exposure to dandelion was at about age seven when my father would pay me a nickel to dig them out of our front yard lawn and throw them into the trash. Boy, things have changed! These days, I would not consider having a front lawn, and I definitely would not dig out the dandelions and toss them in the trash.

A young dandelion root

Dandelion is another versatile wild food. It's not native to our area but is now found all over the world. The yellow flowers make the plant conspicuous in fields and lawns, though it's really the leaves and roots that are most used by the forager.

If you want raw dandelion greens, you'll want to pick them as early as possible in the season, or they become bitter. The bitterness is not bad, and it can be mellowed out by adding other greens. Also, oil-rich dressing makes a dandelion salad more palatable.

A cup of dandelion greens contains over 100 percent of the RDA for vitamin A, which (among other things) is good for the skin. A cup also contains over 500 percent of the RDA

A view of dandelion's leafless flower stalks, with one petal per stalk

of vitamin K. It is a great source of many vitamins and minerals, including the B vitamins and lutein. This has led some to call it "poor man's ginseng."

It's understandable that dandelions have gotten more popular—they are, after all, the richest source of beta carotene, even more so than carrots. However, not all the greens sold as "dandelion" in farmers' markets and supermarkets are the genuine leaf. Frequently, we will see various endive relatives sold and called "dandelion."

Young dandelion greens emerging from a crack in an Alexandria sidewalk HELEN W. NYERGES

The roots are also edible. The younger roots, and plants growing in rich soil, are more tender and more desirable. But I have eaten old roots and tough roots and have found a way to make them palatable. Generally, I scrub the roots to get rid of all the soil, and then boil until tender. You can boil them whole or slice them, and when tender, use them in stews and soups.

For a "coffee-substitute" beverage, wash and then dry the roots. Though there are a few ways you can do this, I generally do a coarse grind, and then roast them in the oven until they are mildly brown. Then I do a fine grind and percolate them into a beverage. You can drink it "black" or add honey and cream.

SALSIFY
Tragopogon spp.

A view of salsify buds, flowers, and the seeding head RICK ADAMS

The yellow flower of salsify, Tragopogon dubius MARK VORDERBRUGGEN

There are about 45 species of *Tragopogon* worldwide, and three have been identified in this area, having either yellow or purple flowers. All are introduced.

Use: The root is used most commonly; the tender leaves can also be eaten

Range: Widespread in lower elevations, along roadsides, hillsides, and in developed areas

Similarity to toxic species: None

Best time: Roots are best gathered in the spring; greens can be collected anytime but are best in the spring

Status: Relatively common

Tools needed: Shovel

Properties

This is a fairly widespread and easy-to-recognize plant. Most folks initially notice the large dandelion-like flower, except it is on a much taller stalk, perhaps two feet or so tall. Depending on the species, the flower may be yellow or purple, and it will have noticeable bracts that extend beyond the petals. The yellow flower is typically *T. dubius*. The purple flower is *T. porrifolius.*

The seed head is just like the dandelion seed head but bigger and very round, around four inches in diameter. The leaves are linear, almost grass-like, and will exude a milky sap (like dandelion or sow thistle) when broken.

Salsify is a biennial plant, meaning that it produces leaves in the first year, and in the second year it sends up its flower stalk before it dies. The roots of the first-year plant are the most tender. If you pick the root from a flowering plant, it will be tough and a bit bitter. These are sometimes cultivated, and in those cases, they are likely to produce larger and more tender roots. But in the wild, the roots you're likely to find are thin, like pencils.

Uses

Salsify, also called the oyster plant, is most often used for the root. Before I ever ate one, I read the descriptions of "large fleshy roots," with an oyster flavor, and imagined something like a carrot or radish that would cook up into some exotic seafood-like dish. However, the reality is a little different. Most of the time, I find the plant growing in compacted and hard soil, and so the root never had a chance to get large and fleshy. It is more like a slender carrot, a bit fibrous, and though the flavor is somewhat bland, the texture is improved by cooking.

Of course, it didn't help that when I was first digging up the salsify roots, they were all from the flowering plants because those were the only ones that I was able to recognize as salsify. But these were the older, second-year plants with tougher roots.

Yes, there are cases where you will find an easy-to-dig root, which is pretty darn good, such as during the rainy season in mulchy soil. Either way, this is such a common plant that you should at least try it. And because it's an introduced exotic, no one will mind if you do some volunteer weed removal.

A view of the first-year root of salsify, before the plant has flowered

Try cooking it to tenderize, and then slice the root and add to soups and stews. In general, you'll need to cook these roots about two to three times as long as you'd need to cook potatoes. Does it really taste like an oyster? Maybe, maybe not. Perhaps there is a slight flavor reminiscent of something like oyster, but that's not the way I'd describe it. Clearly, it has a unique flavor and texture; I find it reminiscent of gobo/burdock root.

We have also cooked up all of the aboveground plants and eaten all that was tender. The stalk will usually be tough, but the leaves are tasty and chewy.

Botanist William Schlegel reports that his favorite part of the salsify is

A view of the first-year leaves of the salsify plant

the flower buds of *T. dubius*. He uses them as a nibble when he's out on his land. At the bud stage, these are tender and make a mild vegetable.

When doing underground pit cooking, we have covered potatoes and onions and meat with the upper part of the salsify plant—the leaves and tender stems—which were abundant. Not only did these leaves protect the vegetables we were steaming, but we found these greens to be a tasty addition to our meal as well.

LORE

Sometimes, salsify is known as "John goes to bed at noon." The flowers begin to slowly close around 11 a.m. and are closed by noon. The flowers stay closed for two to three hours and then open around 3 p.m., and then they stay open until dark when they close again. These plants were sometimes used in European "clock gardens."

GROUP 9

COMMON SUNFLOWER
Helianthus annus

The wild sunflower thrives in this front yard. LILY JANE TSONG

There are 67 known species of Helianthus, with at least 18 species in this area.

Use: Mostly seed for food

Range: Gardens, roadsides, fields

Similarity to toxic species: None

Best time: Summer

Status: Widespread

Tools needed: A collecting bag

Properties

You've no doubt seen flowers in fields many times and thought, "Hey, there's a sunflower." Maybe it was; maybe it wasn't. After all, the Sunflower *family* is one of the largest botanical families. If what you saw was tall and looked very much like the sunflower plants that we grow in our gardens but perhaps just a bit smaller, then you've likely encountered the actual sunflower plant.

The sunflower plant can grow up to eight feet tall, with its erect stem, usually unbranched but sometimes branched. It's covered in stiff hairs. The leaves are alternately arranged and are generally triangular in shape; each leaf is more or less heart shaped and narrowing to a tip. The leaves are also covered in stiff hairs.

It's the flower that is universally recognized. The flower heads are around six inches across, with yellow ray flowers ("petals"). The flowers of the central disk have no showy petals and are brown. The fruits are single-seeded, dry, and flat.

The plant is common in disturbed soils, gardens, roadsides, fields.

Uses

The seeds are the primary food, though sometimes the wild sunflowers do not produce abundant volumes of seed to make gathering worthwhile, or the seeds can be very small and dry. If there is a sufficient volume for collecting, they are collected when ripe and shelled. The seeds are then used in bread products and pastries, desserts, drinks, and as snacks. The shelled seeds can be used raw or roasted, in any recipes calling for nuts, including breads and cakes. Finely ground sunflower seeds (shell and all when they are tiny) can be used as a partial substitute for flour in some recipes.

Native people would roast and grind the entire seeds into a fine flour. The flour would then be mixed with water and made into a drink or made into a dough and mixed with other ingredients.

Botanist William Schlegel told me about the classic book *Buffalo Bird Woman's Garden*, where the author talks about how hard it is to grind sunflower seeds, and so she grinds them shell and all. Schlegel pointed out that sunflower seeds can be hard to shell even with modern equipment. "Imagine the fiber!" said Schlegel.

One method to separate shell from seed is to coarsely break the seeds and pour them all into water. The seeds will sink, and the shells will float and then can be removed. Those shells can be dried and roasted and used to make a coffee substitute, often mixed with other ingredients.

A view of one flower and immature buds BARBARA EISENSTEIN

Immature flower buds or heads can be boiled and eaten.

The wild sunflower seed doesn't make as good a nut butter as the cultivated sunflower seed. However, you can still grind the shelled seeds and mix them with honey, butter, or various oils to make a wild sunflower butter.

According to naturalist Malcolm McNeil, "Sunflowers were one of the common crops grown in fields by Native Americans. The tall flowers provided food, but they also served as support for other plants being grown. Beans would be planted at the base of the flowers and wind their way up the tall stalk. I have tested this in my own garden and can attest that they are effective growing poles for beans."

TALES FROM OVER IN APPALACHIA: Linda Sheer from rural Kentucky was a friend and mentor. She told me that when her people didn't have enough "regular" food, the young, immature flower heads of the sunflower would be collected and boiled until tender and eaten. She prepared and served these to us at one of our cooking classes. Though still a bit chewy and tough, they were certainly edible and palatable, and her unique seasonings made them tasty. When I find these in the spring, I still collect them, cook them, and enjoy them with butter, reminding myself that just about any food is good and satisfying if you haven't had anything to eat in three days.

GROUP 12

COLTSFOOT
Tussilago farfara

A view of a coltsfoot patch HELEN W. NYERGES

Use: Leaves used medicinally and as a non-nicotine smoke
Range: Found widely throughout the area, in mountainous and forested areas, along road-sides, more wilderness than urban areas

Similarity to toxic species: None
Best time: Spring into summer
Status: Relatively common
Tools needed: Collection bag

Properties

Coltsfoot is a perennial herbaceous plant that spreads by its underground rhizome and also reproduces by seed. The flowers first rise on long stems covered in scale-like leaves. The flowers are yellow and somewhat resemble dandelion flowers, with an outer row of bracts.

The large leaves appear after the flowers have seeded, withered, and died. The common name comes from the leaf's supposed resemblance to a colt's foot. The leaf might be up to eight or nine inches across, with a conspicuous white underside. The leaves have angular teeth on their margins.

Overall, the plant grows about 10 inches or so in height, often in thick colonies.

The plant is somewhat widespread throughout the world, especially in Asia, Northern Africa, and Europe. It is a common plant in North and South America where it has been introduced, most likely by settlers, as a medicinal item, probably for coughs. One of its other names is coughwort in Old English. The plant is often found in waste and disturbed places and along roadsides and paths.

Christopher examines coltsfoot leaves in a patch growing along a trail in Virginia. HELEN W. NYERGES

A cluster of coltsfoot plants. Note the overall round shape of the leaf and the deep cleft where the stem meets the leaf. HELEN W. NYERGES

Uses

Though coltsfoot leaves *could* be cooked and eaten, they are really marginal in terms of a food item.

The aromatic leaves have long been infused and used as a tea or to relieve coughing. The leaves have also been boiled and the water used to flavor candy or other confections, even medicines.

The leaves can be dried and smoked, either as a remedy for mild asthmatic conditions or just because one wants to smoke something safer than tobacco.

In my early years of botany studies, this is one of the first herbs of the East that I learned about when I lived on my grandfather's farm, where it grew on the edges of clearings in the woods. I took the leaves and smoked them in my pipe, often mixing mint or other mild herbs.

I have dried and powdered the leaves, and I have used them for a seasoning and found them subtle, mild, and certainly useful if I had no other seasonings.

Also, following the advice of others, I dried and burned the leaves and used the ash as a seasoning. This ash is often described as a salt substitute. I would not describe it as a salt substitute because the flavor is different than salt. However, it did impart a unique flavor to my egg dishes, salads, and some soups that I found acceptable to good.

GROUP 13

GOLDENROD
Solidago spp.

Goldenrod flowers

There are about 150 species of *Solidago* in North America; there are at least 28 species in our area.

Use: Leaves made into a tea; used medicinally

Range: Found widely throughout the area, often in fields and open areas; widespread

Similarity to toxic species: None

Best time: Spring through summer

Status: Common

Tools needed: Collection bag

Properties

Goldenrod is a perennial herb with leafy stems that rise from the rhizome. The elongated, lance-shaped leaves are alternately arranged and attached directly to the main stalk without a petiole (leaf stalk). The leaf margins can be entire or toothed. Many of the goldenrods look very similar, and most have teeth or serrations on the margin. The preferred goldenrod, *Solidago odora*, has an entire margin—that is, there are no teeth at all on the margin. When you crush the leaves of *S. odora*, you will smell a sweet, anise-like aroma. When you hold a leaf up to the sunlight, you will observe that the leaf is dotted with translucent spots. Regardless of species, all goldenrods can be used for a tea.

The unbranched flower stalk can rise up a few feet and is very conspicuous. The yellow radiate flower heads are small and numerous, formed along the top surfaces of the long branching and slightly recurved stems.

Uses

The young leaves of goldenrod have been used as a potherb, cooked and used like spinach. Donald Kirk, the author of *Wild Edible Plants*, says that it makes a good potherb. Of the

A view of the young goldenrod leaf MALCOLM MCNEIL

60-plus species of goldenrod found in the United States, only *S. odora* is suitable for food, according to Alan Hall (author of *The Wild Food Trailguide*).

Goldenrod is apparently not *commonly* used for food, because you find a complete absence of goldenrod leaves mentioned in any recipes. This is probably because the leaves are a bit aromatic and sometimes tough. Linda Sheer, who grew up in rural Kentucky, used to tell me that women on the farm (which actually meant her family living deep in the hills) could make just about any wild green palatable with chopping and cooking and good seasoning. Goldenrod leaves might fall into this category, as it is a green that is not commonly used for food, but certainly could be, especially when there was nothing else available.

More often, the dried leaves and mature flowers of goldenrod are used for a flavorful tea.

The goldenrod plant before it begins producing its flowers MALCOLM MCNEIL

Though we have tried tea from the fresh leaves, drying seems to produce an improved flavor, and drying is nearly always suggested by everyone who discusses using goldenrod leaves for tea. The tea is made by infusion.

Other Uses

Periodically, you will see a swollen gall on a goldenrod stem. The entire dried gall can be used as a float for fishing. If the gall is fresh, you can cut it open and maybe you will find small larvae, which can be used for fishing bait.

The straight stalks of goldenrod have been used for hand drills for making fire.

Cautions

Some individuals have experienced allergic reactions to goldenrod.

A view of the goldenrod flowers KATHLEEN ASHLEY

The flowering goldenrod plant VICKIE SHUFER

Goldenrod growing along a Virginia trail

BARBERRY FAMILY (BERBERIDACEAE)

The Barberry family consists of about 16 genera and approximately 670 species worldwide. Only *Berberis vulgaris* is found in this area.

The mature fruit of the barberry

COMMON BARBERRY
Berberis vulgaris

The fruits of all members of the *Berberis* genus are edible. There are approximately 600 species of *Berberis* worldwide, which are widespread in North America.

Use: Fruits eaten raw or made into wine, jams, or jellies

Range: Found widely throughout the area; common in the mountainous and forested areas; sometimes used as an ornamental

Similarity to toxic species: None

Best time: Summer

Status: Common

Tools needed: Berry-collecting basket

Properties

This barberry is a deciduous shrub, which can grow a dozen feet tall. The leaves are clustered in groups of two to five, with small three-branched spines.

The yellow flowers measure about a quarter-inch wide, produced on panicles in the spring. The conspicuous fruit is an oblong red berry about a half-inch long by about a quarter-inch wide. These mature in the late summer into the fall.

Though barberry can be found throughout the area, it is most common in wooded areas. It is also planted as an ornamental, so it will be found outside its native terrain.

Uses

My experience with the barberry is mostly as a trail nibble, and on occasion, I have had a least a handful to mash and use as a pancake topping. These small oval berries are tart and refreshing, high in vitamin C, and make a good jelly when lightly sweetened with honey.

The vitamin C–rich berries have a distinct flavor, and are an excellent berry, though they seem to not be eaten as much as other readily harvestable fruits. Clearly, many birds consume the fruits and spread the seeds.

The barberry fruits—with their high pectin content—have long been used in Europe and southwestern Asia (especially Iran) for making jams. In Iran, the fruits are added to rice pilaf dishes.

According to Cecilia Garcia and Dr. James Adams in *Healing with Medicinal Plants*, the fruits of all the members of the *Berberis* genus were eaten raw or cooked by most Indigenous peoples of North America, wherever the plant grew.

Sometimes the fruits were dried, then ground into a flour that was used for a mush. Many of the Indigenous people made drinks from these fruits. They are also quite good dried and used as a snack food or added to cookies, cakes, or other dishes as you'd add raisins.

The flowering barberry plant

BIRCH FAMILY (BETULACEAE)

The Birch family consists of six genera and about 155 species worldwide.

A view of the fruit MARGO BORS

HAZELNUT, AKA BEAKED HAZELNUT

The *Corylus* genus has about 15 species worldwide, with only *C. cornuta* and *C. avellana* found here.

Use: Edible nuts

Range: Found on the edges of forests, slopes, and other shady habitats

Similarity to toxic species: None

Best time: Late summer into fall, when nuts mature

Status: Common

Tools needed: Bag or box for collecting

Properties

This large shrub, or small tree, can grow up to 25 feet tall and can easily be confused for an alder. The leaves are oval to round, alternate, with a rounded base and pointed tip. The whole leaf is about three inches long, with double-toothed margins. The leaves are more or less hairy (actually it's better described as a fine fuzz) on both sides.

The nuts, which mature from September through October, are formed in pairs. A papery, bristly, outer husk covers the nut, which has a thin, brittle inner shell. When you see the exposed nut, it will remind you of a commercial filbert, to which it's related. It's relatively easy to identify this tree when you find it and easy to harvest the nuts.

The nuts ripen in the summer and autumn and remain clinging on the trees until they are picked or shaken free by wind or animals.

Uses

This is an excellent nut, and you'd use it in any of the ways in which you'd use a filbert: raw, roasted, slivered, and so on. This means you can shell them and eat them raw in nut mixes, in salads, and even sprinkled into bread or pancake batter. Try sprinkling them on ice cream. The nuts can also be ground into a meal and used to form cakes or added to other pastry dough.

A view of the hazelnut tree MARGO BORS

Nuts are one of the great survival foods since they have the oils necessary for life. If you harvest enough to store for later use, the easiest preservation method is to let them dry (in the sun, or in an oven at pilot light temperature). Then they can be packed into glass, metal, or plastic containers (whatever you have that can keep the mice and rats out), and stored with some dessicant packs. They store a long time, provide some quick energy with no cooking, and give our bodies a lot of what they need.

A view of the leaf and the paired fruit of the hazelnut KEIR MORSE

MUSTARD FAMILY (BRASSICACEAE)

The Mustard family is another large family, comprising more than 330 genera worldwide and about 3,780 species. This large family is subdivided into eight groups.

The floral characteristics that define the Mustard family are that it has four free petals, four sepals (generally white or yellow but other colors as well), six stamens (four long and two short), one pistil, a superior ovary, and fruits—generally a capsule or silique with two valves. Many are cultivated for foods and some for ornamentals.

My mentor, Dr. Leonid Enari, stated that he was unaware of any toxic member of this entire family, though some are more palatable than others. As a result, I have experimented with many of the Mustard family species in various parts of North America. The most obvious edible members are presented here.

The typical leaf shape for this family is lyrately pinnate, meaning a large terminal lobe and smaller lateral lobes. Once I asked a fellow botanist for help in identifying the genus of a plant I'd found, which I knew was in the Mustard family. He replied, "Trying to identify a *Brassicacea* when not in flower is not exactly fun." Well said! These are much easier to identify once they have flowered.

Mustard leaves and stems

MUSTARD
Brassica spp.

There are 35 species worldwide, and many are found throughout the United States. At least six species are found in this area.

Use: Leaves raw or cooked; seeds for spice; flowers for garnish

Range: Fields, urban areas, lowlands, hillsides

Similarity to toxic species: None

Best time: Spring for greens and flowers

Status: Widespread

Tools needed: None

Properties

Though you should learn to recognize the common mustards even when the plant is not in flower, it is the flower that will initially draw you to the plant. The bright yellow flower has the typical Mustard family flora arrangement: four petals (shaped in an X or cross), four sepals (one under each petal), six stamens (four long, and two short), and one pistil. These

A view of the mustard leaf and rosette

Barbara Kolander collecting mustard flowers to use in her wild food cooking workshops

FORAGER NOTE: This is a hardy plant, often outliving many other greens as the season progresses. I have managed to find some mustard greens even during droughts when no other greens were available.

A mustard rosette

are formed in a raceme with the buds toward the tops, then the mature flowers, and then, lower on the stalk, the seedpods forming. The seedpods are about an inch long and needle-thin.

The initial basal leaves are lyrately pinnately divided, meaning that they have the appearance of a guitar with a large round lateral lobe and smaller side lobes. Not exactly like a guitar, but that gives you a good mental picture. As the plant matures, the leaves that form on the upper stalks are smaller and linear and look nothing like the young basal leaves.

Uses

Mustard is one of the very first wild foods that I began to eat, partly because it is so common and partly because it is so easy to identify. I recall seeing a line drawing of it in Bradford Angier's book *Free for the Eating*, which didn't look anything

RECIPE

Pascal's Mustard

Fellow forager Pascal Baudar takes the pungent flowers of regular black mustard and grinds them while fresh, adding white wine and vinegar to taste. He thus produces a mustard condiment from the flowers, not the seeds as is the usual custom. This makes a delicious mustard with a new twist.

like the green plant with yellow flowers that I was seeing everywhere. Angier used a picture of the mature plant gone to seed, and I saw the young spring plant. They were both right, but it demonstrated the need to always learn plants by observing them in the field.

I began with the young mustard greens, chewing the raw leaves and enjoying the spicy flavor, despite the fine hairs covering the leaves (not all *Brassica* are hairy). I then moved on to chopping them up and adding to salads, which was good. I then began to boil the leaves and serve them to my family with butter. Everyone enjoyed them, even my father. Eventually, I found that I could add mustard greens to just about any dish: soups, mixed salads, omelets, stir-fries, potatoes, you name it!

The flower buds and flowers have also been a good trail treat and make a good colorful garnish for salads and soups. I give them to children and tell them that they taste like broccoli, and most of the children say they enjoy the flowers.

The tender tops of the stems with the flower buds can also be snapped off the upper part of the plant, steamed, and served with some sauce or cheese. The flavor is just like the Chinese broccoli that you buy at farmers' markets.

Lastly, you can come back to this annual plant in late fall, when the leaves are dried up and the tops are just tan-colored stems with small seedpods. Collect the pods in a bag and break them up. The seeds go to the bottom of the bag, and you can discard the pod shells. The brown seeds are then used as a seasoning for various dishes calling for mustard, or you can try making your own mustard from them.

SEA ROCKET
Cakile edentula and *C. maritima*

Sea rocket in flower, growing on the beach

Use: Greens, sprouts, and flowers; ideally cooked but can be used raw sparingly

Range: Restricted to the sandy beaches along the coast

Similarity to toxic species: None

Best time: Spring, but can be picked year-round

Status: Somewhat common

Tools needed: None

Properties

There are seven species of *Cakile* worldwide, found on the beach shores of North America, Africa, and Europe. In our coastal areas, we have *Cakile edentula* and *C. maritima*.

Sea rocket is found along the coast, growing in the sand in the upper areas of the beach, usually just beyond high tide in the dunes. When you see how well these plants have naturalized, it is hard to believe that they are not natives. *C. edentula* is the one we're most likely to find on our beaches.

The leaves are very much like a small mustard leaf but plump, as if they were swollen. Each leaf has a bluish-green appearance, with the leaves pinnately divided into linear segments. Typically, each leaf tends to fold inward along the central vein of the leaf.

The seedpods also look swollen, and you can see why the plant is called "rocket" by looking at the pod's space-rocket appearance.

Of course, the lavender to light-purple flower has the typical mustard flower arrangement of four petals, four sepals, six stamens, and one pistil.

An individual sea rocket leaf RICK ADAMS

Young sea rocket leaves RICK ADAMS

Uses

The leaves are strongly flavored like horseradish, and generally, you would not want to include the mature leaves in a salad. But boiling tones them down quite a bit so they are then tastier and more palatable. The boiled leaves can be added for flavor to soup broths or to dishes of mixed greens. In general, you'd probably not want to serve them alone as a cooked green unless you changed the water once and then served them with some onions and probably a savory sauce.

Still, they can turn an otherwise bland meal into quite a treat. They will help to flavor clam chowder as well as other soups and stews. They will really liven up stale old MREs (Meals Ready to Eat).

RECIPE

Wild Wasabi

We have finely diced the sea rocket leaves into nearly a paste and added and mixed in a very small amount of oil and vinegar, creating a passable "wasabi."

Once, while visiting a coastal area, my friends and I were hiking in the sand above the high tide line. It was very foggy in the wintertime, and most of the sea rocket was dried and gone to seed. Under one very large old bushy sea rocket plant, we noticed hundreds of the little sprouts in the sand. We carefully harvested many of these. Each was tender and not at all harshly flavored, as the more mature plant tends to be. We added some to salad for our mostly foraged meal that evening, and the rest we added to our homemade clam chowder.

Think of sea rocket more as a flavoring agent and garnish, not as a principal food.

A view of sea rocket seedpods RICK ADAMS

SHEPHERD'S PURSE
Capsella bursa-pastoris

Shepherd's purse leaves

There are four species of *Capsella* worldwide.

Use: Leaves eaten raw or cooked; medicine

Range: Prefers lawns, fields, and disturbed soils

Similarity to toxic species: None

Best time: Spring is best for greens; the seeds can be collected late spring to early summer

Status: Somewhat common

Tools needed: None

Properties

Shepherd's purse, which is found all throughout the United States, is most easily identified by its flat, heart-shaped seedpods. They are unmistakable! The stalks rise about a foot or so tall. The little clusters of white flowers, sometimes tinged with a bit of purple, are formed in racemes along the stalk. These then mature into heart-shaped pods. The trouble is, by the time you see all the seedpods, it's usually too late to use the young leaves for food, but now you know how to recognize shepherd's purse for the next season.

The young leaves are often hidden in the grass, making them somewhat inconspicuous. The basal leaves are toothed, with a large terminal lobe, typical of Mustard family leaves. The upper leaves are without a stalk and are more arrowhead shaped. If you look closely, the young leaves will be covered with little hairs.

A view of the heart-shaped seed pods of shepherd's purse

Uses

The flavor of shepherd's purse leaves is mild, and they could be used in just about any recipe, such as salads, sandwiches, soups, eggs, and so on. However, they seem to be best when used in salads. Additionally, once Europeans had significantly flowed into North America after the first East Coast colonies, shepherd's purse became so common that the Indigenous people began to collect and grind the seeds into a meal. The ground seeds were then used in drinks and as a flour for various dishes.

Dr. Leonid Enari used to poll his students on which plant tasted the best of the many wild plants he let them try. Consistently in his polls, shepherd's purse was rated the best. It is actually somewhat bland and peppery, but not *too* peppery, and the texture is mild. Even finicky eaters will like these leaves.

It's also very nutritious. About a half-cup of the leaves (100 g) contains 208 mg of calcium, 86 mg of phosphorus, 40 mg of sodium, 394 mg of potassium, 36 mg of vitamin C, and 1,554 IU of vitamin A.

If you don't know this plant yet, you're most likely to recognize it by

A view of the seeding stalk with the heart-shaped seed pods

the heart-shaped seed pods, which means it's late in the season and probably too late to use any of the greens. So, come back to that spot in the following season so you learn to recognize the very young leaves.

Dr. Enari told his students that this was the best plant to stop nosebleeds. You boil the plant, dip a cotton ball into the water, and then apply to the nose. It turns out that many people have used this plant medicinally, especially to stop internal or external bleeding.

WATERCRESS
Nasturtium officinale

A view of watercress growing at the edge of a stream

There are five species of *Nasturtium* worldwide.

Use: Leaves eaten raw or cooked in salads, stir-fries, soup, etc.; can be dried for use as seasoning

Range: Restricted to the edges of lakes and streams

Similarity to toxic species: None

Best time: Summer, before the plant flowers; however, the plant can be collected anytime

Status: Somewhat common along streams

Tools needed: None

Properties
Watercress is found worldwide along slow-moving streams and lakes and is common in North America.

Once you learn to recognize watercress and see how the pinnately divided leaves are formed, you will find it quite easy to recognize, whether it is very young or older and flowering. First, it nearly always grows directly on the edges of streams where the water is slower. Occasionally you'll find it in sandy areas, but it is always an area that is at least seasonally underwater. You'll typically find it growing in thick mats.

Watercress in flower HELEN W. NYERGES

The leaves are pinnately divided into round leaflets. The stems are hollow, and there are white hairs on the underwater part of the stem. The plant is in the Mustard family, so when it gets older and flowers, the white flowers will be divided into the typical Mustard formula: four petals, four sepals, six stamens, and one pistil.

Though watercress can today be found worldwide, it is regarded by botanists as a native plant. It was known to be a part of the diet of early Native Americans.

Uses

Watercress was one of the very first wild plants that I learned how to identify and began to use. It is not only common throughout waterways in this area but throughout the world.

I have always enjoyed making a salad of mixed greens and including watercress. But I don't usually make a salad with *only* watercress because it's a bit too spicy for my taste. A few raw watercress leaves are also tasty in sandwiches.

Watercress makes a delicious soup. Just finely chop the entire plant (tender stems and leaves) and add it to a water- or milk-based soup. Or you can add chopped watercress leaves to a miso-based soup.

You can cook the greens like spinach, serving it with a simple seasoning such as butter or cheese. Or, try mixing the greens into an egg omelet. If

Christopher collecting watercress BARBARA KOLANDER

you're living off MREs or freeze-dried camping food, you can add some diced watercress to liven up your meals.

Also, for those of you who like making your own spices, you can dry and powder watercress and use it to season various dishes. Use it alone or blend the powdered watercress with powdered seaweed or other flavorful herbs. You'll notice that some of the commercial salt-alternative spices use dried watercress leaves.

Watercress

Another self-reliance idea is to dry wild foods into the basis of soup stock and then reconstitute later into a soup or stew broth. Dried and powdered watercress makes an ideal ingredient in such a mix.

Watercress has numerous medicinal applications. Some of the most popular, and most documented, include eating watercress to prevent eczema from returning or for inflammatory flare-ups, as well as using an extract of watercress (obtained by boiling the leaves for 10 minutes in water) as a disinfectant to wash the eczema area.

As a tea, watercress acts as a digestive aid, as the mustard oil glycosides, vitamins, and bitters in the tea promote the production of the stomach juices that aid digestion. With its high potassium content, watercress tea assists the kidneys by acting as a diuretic, cleansing the urinary tract.

Cautions

If you have doubts about the purity of the water where you get your watercress, you should not eat it raw; boil it first and then use it in a cooked dish. Always wash the watercress before using it. It grows right in the water, and you want to remove any dirt or other undesirables that may be clinging to the plant.

RECIPE

Saturday Night Special

Gently sauté half an onion bulb, diced, in a skillet with butter. You could substitute a handful of wild or garden onion greens. Quickly add at least 1 cup of watercress, chopped into large pieces, and cook gently until all is tender. Add a dash of soy sauce and serve.

HEDGE MUSTARD AKA TUMBLE MUSTARD
Sisymbrium altissimum and *S. officinale*

A rosette of the young hedge mustard leaves

There are 41 species of *Sisymbrium* worldwide, with at least two found in our area. All are native to Europe. These two are *S. altissimum* and *S. officinale*.

Use: Leaves, raw or cooked

Range: Prefers disturbed soils of fields, farms, and gardens, and along roadsides and trails

Similarity to toxic species: None

Best time: Spring

Status: Common

Tools needed: None

Properties

If you already know the mustards (*Brassica* spp.), you will very likely think "mustard" when you see hedge mustard. The flowers of *Sisymbrium* tend to be smaller than the *Brassica* flowers, and the leaves tend to be pointier compared to the rounder leaves of *Brassica*. Of course, to botanists, the distinction is mostly in the details of the flowers, but with sufficient observation, you'll be able to recognize the hedge mustards by leaf alone.

Uses

I think of the *Sisymbriums* as wild wasabi. Chew on a bit of the leaf, and you'll get that hot horseradish-like effect that opens your nostrils. I have friends who actually turn these leaves into wild wasabi, which is great on sandwiches and crackers or as a dip. But generally, I regard the hedge mustards as a source of very spicy greens that go well with salads, soups, egg dishes, sandwiches, stir-fries—just about any dish where you can add greens. These are spicy greens; in general, they are a bit spicier than the greens of the common mustards (*Brassicas*).

I have had broths made from the finely diced hedge mustard leaves, into which a lot of rice had been

Someone else also likes the hedge mustard leaves.

RECIPE

Screaming At The Moon

A hedge mustard soup recipe.

2 cups chopped young hedge mustard greens

2 garlic cloves, peeled and crushed

3 cups water

¼ cup miso powder

Simmer the greens and garlic with the water in a covered pot. When tender, add the miso and cook for another 5 minutes. Serves 2 to 3.

A hedge mustard rosette

added. This dish was hot and good! I have also had "wild kimchi" that consisted of wild greens that had been marinated in raw apple cider vinegar. A lot of hedge mustard leaves were used in one of these kimchis, and it was delicious. You could also dry the hedge mustard leaves and either reconstitute later or just powder them and use as a seasoning.

The flowers are good too, but they aren't quite as good as the *Brassica* mustard flowers. Hedge mustard flowers seem to have too much of that bitter and astringent bite, so I use them sparingly in soups, salads, or other dishes.

CACTUS FAMILY (CACTACEAE)

There are 125 genera in the Cactus family, and there are about 1,800 species worldwide, mostly in the American deserts. In our area, though, you might find various species in protected gardens, but we generally will only find *Opuntia humifusa* in the wild.

The youngest pads are easiest to clean for eating.

PRICKLY PEAR
Opuntia humifusa

Use: Young pads for food, raw or cooked; fruits for deserts and juices; seeds for flour
Range: Opuntias are found all over North America and Southern Europe. In the United States, they are most common in the Southwest. In our area, they are found sporadically in the lowlands.
Similarity to toxic species: None
Best time: Spring is the best time to collect the new pads, though the older ones can also be used. September through October is the best time for harvesting the fruits.
Status: Not common
Tools needed: Metal tongs, sturdy bucket, possibly gloves

Properties

The prickly pear cacti are readily recognized by their flat oval pads, with their spines evenly spaced over their surface. The cacti flower by summer, and then the fruits mature by August and September.

The fruits tend to have more spines, so I collect those with metal tongs. Still using the tongs, I turn each fruit over in a flame—about 10 seconds—to burn off the spines and glochids. Then I cut them in half, remove the fruit inside, and eat, preserve, or process in some way.

Prickly pear cactus growing in a Virginia yard
ALEX CHAN

Uses

There are better cacti for eating than the species found here, but if you're stuck somewhere where cactus grows, one of these might comprise a part of your meal. And there are several ways to get a meal from the prickly pear cactus: young pad, old pad, fruit, and seed.

The new growth of spring offers one of the more readily available foods with the least amount of work. Remember, the cacti all have some spines and tiny little glochids, so you'll need to be careful whether you have the very young or the very old pads.

When you get the very young pads of spring, they are still bright green, and the tough outer layer won't yet have developed. Carefully pick and then burn off the young spines, or thoroughly scrape each side to remove all spines and glochids. Then you can slice or dice and sauté to remove much of the liquid and sliminess of the cacti. Cook off the water, and then add eggs, potatoes, or even tofu for a delicious stew.

Monica Montoya tastes a bit of the inner tender section of an older cactus pad.

FORAGER'S NOTE: If you collect cacti, you will—sooner or later—get spines and glochids in your skin. Spines are easy to see and relatively easy to remove. But glochids are smaller and hair-like and more difficult to remove from the skin. Try smearing white glue on the part of your hand that has glochids. Let the glue dry. Now peel it off. This will usually remove most glochids.

Opuntia humifusa, growing in Virginia

Opuntia humifusa, growing in Virginia, putting out young buds

Young green cactus pads, cleaned and sliced. These are now ready for cooking.

Fruits are delicious, too, and are the closest thing to watermelon that you'll find in the wild, aside from the abundance of small seeds. An excellent drink is made by mixing 50 percent of this cactus fruit puree with 50 percent spring water.

Eating the prickly pear pads (raw, cooked, or juiced) has long been considered one way to combat diabetes. For those who don't want to grow, clean, and cook their own nopales, you can now purchase the powder, which you consume in various ways in order to combat diabetes.

Cautions

If you choose to collect these cacti for food, you will get spines and glochids in your skin, eventually. However, if you practice caution, you can keep this inconvenience to a minimum.

Occasionally, people have experienced sickness after eating certain varieties. In some cases, this is due to a negative reaction to the mucilaginous quality. There may be other chemical reasons as well. So, despite this being a very commonly used food historically for millennia, we suggest you start with very little and monitor your reactions.

PINK FAMILY (CARYOPHYLLACEAE)

The Pink family consists of 83 to 89 genera (depending on which authority) and about 3,000 species worldwide.

Fresh chickweed collected, rinsed, and ready for salad

CHICKWEED
Stellaria media

There are 190 species of *Stellaria* worldwide, with at least seven found in our area.

Use: The leaves are best raw in salads but can also be cooked in various dishes or dried and powdered to make into pasta.

Range: Moist and shady areas in urban settings, gardens, along rivers, and shady wooded areas. Scattered widely where the conditions are ideal.

Similarity to toxic species: You may find young common spurge (*Euphorbia peplus*) in chickweed patches, which superficially resembles chickweed, but spurge doesn't have the line of white hairs; its stalk is more erect, and the leaves are alternate, not opposite like chickweed. If you break the stem of spurge, you will see a white sap; it shouldn't be eaten.

Best time: Spring; chickweed rarely lasts beyond midsummer

Status: Common

Tools needed: None

Properties

Chickweed is one of the introduced *Stellarias* that is now spread worldwide and everywhere in North America, including here. It is common in urban yards, gardens, shady fields, and canyons. It is a short-lived annual that shrivels up by summer when the soil is dry.

Chickweed is a low-growing, sprawling annual that first appears after the winter rains. The thin stem will grow up to a foot long, and upon close inspection, you'll see a line of fine white hairs along one side of the stem. The oval-shaped leaves, arranged in pairs along the stem, come to a sharp tip. The flowers are white and five-petaled, though they may appear to have 10 petals because each flower has a deep cleft.

Young chickweed in flower

Uses

Chickweed is probably best used as a salad ingredient. In a thick patch of chickweed, one can cut off a handful of the stems just above the root. Then you just rinse the leaves, dice, and add salad dressing.

The plant can also be cooked in soups and stews. For those who are more adventurous, the entire chickweed plant (aboveground) can be dried, powdered, and mixed fifty-fifty with wheat flour, then run through a pasta machine. The result is a green pasta with a flavor of chickweed.

Because chickweed grows close to the ground with its fine stems, it

Young chickweed

is common to find other plants growing in chickweed patches. So, you need to make certain you are only collecting chickweed. We've seen poison hemlock growing within chickweed patches.

A view of individual chickweed plants

A view of chickweed in the garden

RECIPE

Mia's Chickweed Soup

Although chickweed can be found in city sidewalks, it's best to gather it in the wild, away from pesticides. As an homage to their humble origins, I call this my "Sidewalk Soup." It's simple, low-fat (you can omit the pancetta or bacon, and it's still amazing), and has a surprising depth of flavor reminiscent of spring peas and pea shoots. This is my version of "wild split-pea soup."

4–5 tablespoons diced pancetta (or bacon)

1 medium onion, diced

1 stalk celery, diced

1 carrot, diced

1 teaspoon olive oil, as needed

4–5 cloves garlic, finely minced

1 teaspoon fennel seeds

1 small Oregon myrtle leaf

1 small leaf white sage

2 teaspoons French or Italian herbs (I like oregano, thyme, and parsley)

1 small potato, cubed

6 cups packed chickweed, washed and chopped

1 teaspoon raw apple cider vinegar (to keep mixture green)

Salt and pepper to taste

In a heated stockpot, sauté the pancetta or bacon until crisp. Add onion, celery, and carrots and sauté until translucent. You may need to add a bit of olive oil to the bottom of the pan, approximately 1 teaspoon. Add the garlic and spices and continue to sauté until just fragrant. Add the cubed potato; it will serve to thicken the soup, once pureed. Add the chickweed (save a handful for garnish) and enough water to cover the chickweed with an inch of water. Cover and bring to a boil. Add the vinegar, then reduce to a light simmer for about 20 to 30 minutes.

Once slightly cooled, transfer to a food processor and puree the mixture. Add salt and pepper to taste. Serve with tender, crisp chickweed as garnish. Delish!

—Recipe from Mia Wasilevich

GOOSEFOOT FAMILY (CHENOPODIACEAE)

The Goosefoot family consists of 100 genera and about 1,500 species worldwide, found especially in saline or alkaline soils. Some are cultivated for food. Some botanists have lumped this family into Amaranthaceae.

According to Dr. Leonid Enari, this is one of those very promising plant families for food. His research indicated that most of the leaves could be used for food, either raw or cooked, if too bitter and unpalatable. Dr. Enari also stated that the majority of the seeds could be harvested, winnowed, and ground and used for a flour or flour extender.

Rain beads up on the white lamb's quarter leaves.

LAMB'S QUARTER, WHITE AND GREEN
Chenopodium album and *C. murale*

There are about 100 species of *Chenopo-dium* worldwide, and at least 14 species are recorded here.

Use: Leaves eaten raw or cooked; seeds added to soups or bread batter; leaves dried for seasoning

Range: Prefers disturbed soils of farms, gardens, hillsides, fields, along trails, etc.

Similarity to toxic species: Black night-shade leaves can be confused with lamb's quarter leaves when very young. Be sure to look for the white mealy (and "sparkly") underside of lamb's quarter and for the streak of red in the axils.

Best time: Spring for the leaves; late sum-mer for the seeds

Status: Common and widespread

Tools needed: None

A view of the red that is often found on the stem and axils of lamb's quarter

Properties

Lamb's quarter is a plant that everyone has seen but probably not recognized. It is found all over North America (in fact, all over the world), as well as throughout DC and the surrounding states. It's an annual plant that sprouts up in the spring and summer in fields, gardens, and disturbed soils, and gen-erally it grows about three to four feet tall. (I did record one at 12 feet, but that's an exception.)

The leaf shape is roughly triangu-lar, somewhat resembling a goose's or duck's foot, hence the family name. The color of the stem and leaves are light green, and the axils of the leaves and sometimes the stem is streaked with red. The bottom of each leaf is

The young white lamb's quarter plant, C. album

covered with a mealy substance, causing raindrops to bead up on the leaf.

As the plant matures in the season, inconspicuous green flowers will appear, and seeds will form as the plant dries and withers.

Uses

Lamb's quarter is a versatile plant that can be used in many recipes. The young tender leaves can be cut into smaller pieces and used in a salad. The leaves and tender stems can be cooked like spinach and seasoned for a tasty dish. The water from this cooking makes a delicious broth. The leaves are a versatile green, which I've used as an addition to soups, egg dishes, and quiche, and I have even stir-fried it with other vegetables.

Lamb's quarter will go to seed by late summer, and seeds from the dead plant are harvestable for several months. The seed is an excellent source of calcium, phosphorus, and potassium, according to the USDA. Collect the seeds by hand, place them in a large salad bowl, and then rub them between your hands to remove the chaff. Next, winnow them by letting handfuls drop into the salad bowl as you gently blow off the chaff. The seeds can then be added to soups, rice dishes, and bread batter.

Cautions

Older leaves may cause slight irritation to the throat when eaten raw, without dressing.

Green lamb's quarter, C. murale

Ryan Swank inspects a field of lamb's quarter, C. album.

FORAGER NOTE: Everyone should get to know lamb's quarter. Not only is it widespread throughout the DC and surrounding area, but it can also be found throughout the world. I once spent a week in the mountains eating only lamb's quarter (salad, soup, fried, boiled). It is a plant that I can depend on finding even during a drought when nothing else is available.

Urban farmer Julie Balaa inspects two plants that can be easily confused: black nightshade, on the left, and green lamb's quarter (C. murale), on the right.

RUSSIAN THISTLE
Salsola tragus (formerly S. kali)

A view of the maturing Russian thistle plant. Note the red in the axils of the leaves.

There are about 100 species of *Salsola* worldwide.

Use: The very young new growth can be cooked and eaten

Range: Though seemingly a stereotypical plant of the desert, Russian thistle is somewhat widespread in valleys, fields, disturbed soils, and the fringes of the urban sprawl

Similarity to toxic species: None

Best time: Spring

Status: Can be common locally and seasonally

Tools needed: None

This plant is native to Central Asia and eastern Russia. The United States purchased flaxseed from Russia in 1874, and the Russian thistle seed was in the flax. Russian thistle thus made its first appearance in North America in Scotland, South Dakota.

Properties

Everyone seems to know Russian thistle (aka tumbleweed) from the Westerns: large, dry, round plants that blow across the plains in the wind. And while that is somewhat accurate, there is nothing edible when the plants mature into those large, dry balls.

It is the very young new and tender growth that can be used. The color of the stalk and leaves is pale green, almost a shade of blue, and the leaves are spiny and shaped like needles, maybe one to two inches in length. These leaves are covered with fine hairs, and you can usually observe some red in the axils and on the stalks.

Interestingly, though a very inconspicuous plant, it produces an equally inconspicuous flower that is actually very beautiful if you take the time to observe it. The flowers are small and measure approximately an eighth to a quarter-inch across, and they consist of sepals that appear fragile and paper-like; there are no petals. These flowers are formed individually in the upper axils of the plant.

As the plant matures and gets older, it turns into the dry, round ball up to three feet in diameter, and when its small root is broken free by the wind, it rolls over the countryside and spreads as many as 200,000 winged seeds per plant. No wonder it's everywhere!

Uses

Our main source of food here is the young, tender leaves. Collect them individually so you know they are still

Young shoots of Russian thistle with its needlelike leaves

A view of the mature "tumbleweed" of Russian thistle

A closer view of the Russian thistle plant

tender. Usually, I simply boil and serve with butter. They can also be served plain or with cheese.

Once boiled, taste the juice. It's actually a pretty good broth. You can drink it plain or use it as a soup base.

If I am mixing Russian thistle with other vegetables or greens, I will chop it up a bit first. If the leaves are tender enough, they could be boiled, then mixed into a casserole or even a meatloaf-type dish.

Sometimes, if you get the very young shoots of the Russian thistle, you can quickly dip them into boiling water to reduce their fibrous surfaces. They can then be used in raw dishes without experiencing any irritation to the throat.

RECIPE

Tumbled Rice

Cook 1 cup of rice (use some good rice, like wild rice or long-grained brown rice) according to the package instructions. Separately cook about 2 cups of tender Russian thistle leaves and about 1 cup of hedge mustard leaves. When all is cooked, blend them together while still hot. Give it a bit of garlic powder to taste, and top with shredded Jack cheese before serving.

HEATH FAMILY (ERICACEAE)

The Heath family contains about 100 genera and 3,000 species worldwide.

Blueberry fruit

HUCKLEBERRY AND BLUEBERRY
Vaccinium spp.

The *Vaccinium* genus includes more than 400 species worldwide, with at least five recorded in our area. In general, *Vacciniums* are often referred to as huckleberries, blueberries, cranberries, and even bilberries, depending on the species. Nearly all are native here, including *V. macrocarpon*, which is the common cranberry, native to eastern North America.

There are five in our area: *V. angustifolium*, commonly called common lowbush blueberry, low sweet blueberry, sweet blueberry, or early low blueberry; *V. corymbosum*, commonly called highbush blueberry or swamp blueberry; *V. macrocarpon*, common cranberry; *V. pallidum*, commonly called lowbush blueberry, late lowbush blueberry, hillside blueberry, or upland blueberry; and *V. stamineum*, commonly called deerberry, squaw huckleberry, or buckberry.

Use: The fruits are edible

Range: *Vacciniums* are forest inhabitants, found mostly in woodland clearings and in the woods themselves, mostly coniferous woods. They like moist and shaded areas and north-facing hills.

Similarity to toxic species: None
Best time: Early spring for flowers; early summer for fruit
Status: Common
Tools needed: Collecting basket

Properties

These are shrubs with alternate ever-green to deciduous leaves, which are broadly lance shaped. The stems are trailing to erect. The flower's petals generally number four to five, with a corolla that is cup or urn shaped. The fruit could be red or blue, larger or smaller, and have flattened ends. Generally, the plants with the most desirable fruits are the smaller shrubs, about three feet tall, with the larger sweet, juicy blue berries measuring about a quarter to a half an inch in diameter.

The blueberry plant MALCOLM MCNEIL

The best way to make sure you have identified the plant is to observe it when the plant is fruiting and then take note of the leaf and stem characteristics.

Uses

The fruits of all *Vacciniums* can be eaten, and some are better than others.

The fruits of all species can be eaten raw or cooked. The flavor of the

Blueberry flowers VICKIE SHUFER

ripe fruit can vary from tart to very sweet. They can be used to makes pies and jellies, cobblers, and preserves. The fruits can also be dried for later use and used to make fruit pemmican.

These fruits were widely used by Indigenous peoples and have only some-what recently been cultivated.

According to the USDA, 100 grams of the raw fruit contains 63 mg of cal-cium, 54 mg of phosphorus, 338 IU of vitamin A, 420 mg of thiamine, and 58 mg of vitamin C.

The dried leaves can be infused to make a tasty and nutritious tea. An infusion of the leaves has been used to treat diabetes, as a result of neomyrtilicine, which helps to reduce blood sugar levels. The infusion has also been used to treat urinary tract infections, a result of the tannin in the leaf.

Nutritionists regard blueberries as one of the most nutrient-dense foods, with 24 percent of the recommended daily intake (RDI) of vitamin C, 36 percent of the RDI of vitamin K, 25 percent of the RDI of manganese, and small amounts of various other nutrients.

Blueberry fruit VICKIE SHUFER

Research shows that eating these fruits is good for your immune system due to the presence of pterostilbene and other substances.

Blueberries are widely regarded as the "king of antioxidant foods." Antioxidants protect your body from free radicals, which are unstable molecules that can damage your cells and contribute to aging and diseases, such as cancer. Blueberries are believed to have one of the highest antioxidant levels of all common fruits and vegetables. The main antioxidant compounds in blueberries belong to a family of polyphenol antioxidants called flavonoids. In one study, 168 people drank 34 ounces (one liter) of

The blueberry plant MALCOLM MCNEIL

a mixed blueberry and apple juice daily. After four weeks, oxidative DNA damage due to free radicals was reduced by 20 percent. The findings were roughly the same whether fresh or dried blueberries were used.

Blueberries also appear to have significant benefits for people with high blood pressure, which is a major risk factor for heart disease. In an eight-week study, obese people who had had a high risk of heart disease noted a 4 to 6 percent reduction in blood pressure after consuming two ounces (50 grams) of blueberries per day.

Blueberries are really sounding like a superfood! Blueberries can help maintain healthy brain function and improve memory. According to animal studies, the antioxidants in blueberries may affect areas of your brain that are essential for intelligence. In one of these studies, nine older adults with mild cognitive impairment consumed blueberry juice every day. After 12 weeks, they experienced improvements in several markers of brain function.

Blueberry (V. uliginosum) JEAN PAWEK

Because of the anthocyanins in blueberries, consuming them is also believed to improve the health of diabetics.

This is just a short summary of many of the health benefits of blueberries and other fruits of the *Vaccinium* genus. The clinical data behind each of these medical claims were reported on Healthline.com and are summarized here.

Huckleberry (V. ovatum) in fruit ZOYA AKULOVA

OAK FAMILY (FAGACEAE)

The Oak family includes seven genera and about 900 species worldwide. There are at least 22 species of the genus *Quercus* in the greater DC area. Once processed, the acorns of all varieties of oaks can be eaten.

A bowl of harvested acorns

OAK TREE
Quercus spp.

Some of the common oaks of this area include pin oak, white oak, willow oak, black oak, chestnut oak, live oak, swamp white oak, and red oak

Use: Acorns used for food once leached; miscellaneous craft and dye uses

Range: The oaks can be found in nearly every environment

Similarity to toxic species: The tannic acid in acorns is considered toxic, but it's so bitter that you'd never eat enough to get sick or cause a problem

Best time: Acorns mature from mid-September to as late as January

Status: Common

Tools needed: A bag

Properties
Some oak trees are deciduous, and some are evergreen, and the leaf shapes vary

At an Acorn Processing class, students learn to shell, grind, leach, and cook with acorns.

from simple to pinnately lobed. Some are bushes, but most are trees.. To know your local oaks, you should go to an arboretum or a plant society walk. Oak trees are very common. They are ubiquitous. You can go to the Mall in Washington, DC, and see many—they are especially showy in October when they turn yellow and red.

The fruit of all oak trees is the acorn, which every child can recognize. Some acorns are fat, some are long and thin, and the caps can vary significantly. Still, the nut set in a scaly cap is universally recognized as the acorn. You should have no trouble recognizing acorns wherever you live.

Uses

The nut from the oak tree is the acorn, and the acorn is a wonderful source of a starchy food. Though I have three separate cookbooks on my bookshelf devoted entirely to using acorns in diverse ways, I generally only use acorns

David Martinez grinds raw acorns on a traditional grinding stone.

for cookies, pancakes, and bread.

In the old days, acorns were an important food for the Indigenous peoples. Acorns were typically collected, dried, and then stored. For use, the acorns were shelled, and the shells discarded. The acorns were ground into a flour and then placed on a sloping rock with a lip at the lower end or some other variation of a colander. Cold or hot water poured over the acorn meal would wash out the tannic acid. The meal was most often used as a thickener in soups and stews, making a type of gravy.

Acorns were such an important food that every tribe had their own way of processing them, so there was a lot of variety in how this would have been done. There was a lot of lore surrounding the types of acorns; the acorn meal sometimes had a religious

A cloth has been placed into a colander, and the raw acorn flour is placed into the cloth. David Martinez pours cold water into the acorn flour, which allows the tannic acid to wash out with the water.

significance and would have been used in various ceremonies, in much the way that corn or corn pollen is often used.

A view of the leaching acorn flour

Today, on the trail or in the kitchen, the neatest and quickest way to process the acorns is to first shell them and then boil them and change the water repeatedly until they are no longer bitter. At that point, while they are still wet, I prefer to process them through a hand crank meat grinder to produce a coarse meal. You can dry the coarse meal, and then grind it even finer in a wheat or coffee grinder. The meal is perfect for any product calling for flour. I typically mix the acorn flour fifty-fifty with wheat or other flour. This is partly for flavor and partly because acorn flour doesn't hold together as well as wheat flour.

Gary Gonzales (right) cooks a batch of acorn pancakes while talking with other students.

The more traditional method of processing first involves shelling the acorns and then grinding them while still raw. I typically do this on a large, flat-rock metate. Then, the meal is put into some sort of primitive colander and water (hot or cold) is poured through it. There were many possible ways to create a colander in the old days; today, I just put a cotton cloth inside a large colander and pour cold water over the acorns. Cold water helps to retain the oil and, therefore, the flavor of the acorn meal. The water takes a while to trickle out, and it may require pouring up to three or four gallons of water through it before the acorn meal is no longer bitter and can be eaten.

RECIPE

Forest Memories

1 cup processed acorn flour (with tannic acid removed)

1 cup wheat or spelt flour

Water

Mix the flour and water to create a suitable dough. Form the dough into small loaves and cook on a soapstone slab.

I have had modern acorn products of chips, pound cake, and pasta. They are delicious. If I had to describe the acorn flavor, I would say that products made with acorn have a subtle graham cracker flavor.

How nutritious are acorns? Of the many species analyzed, the protein content ranges from 3.9 to 6.3 percent, fats range from 4.5 to 18 percent, carbohydrates range from 54.6 to 69 percent. In general, the total proteins, fats, and carbohydrates range from between 76 to 79 percent. Source: Martin A. Baumhoff, *Ecological Determinants of Aboriginal California Populations* (Berkeley: University of California Press, 1936), 162, as modified by Carl Brandt Wolf, *California Wild Tree Crops* (Claremont, CA: Rancho Santa Ana Botanic Garden, 1945), table 1, and William S. Spencer, *Handbook of Biological Data (New York:* W. B. Saunders Co., 1956) table 156.

Nutritional and Medicinal Properties of ACORNS (Source USDA)

Vitamins	Amount	DV
Folate	87 mcg	
Folic acid	0 mcg	
Niacin	1.827 mg	9%
Pantothenic acid	0.715 mg	7%
Riboflavin	0.118 mg	7%
Thiamin	0.112 mg	7%
Vitamin A	39.0 IU	1%
Vitamin A, RAE	2.0 mcg	
Vitamin B12	0.00 mcg	0%
Vitamin B6	0.528 mg	26%
Vitamin C	0.0 mg	0%
Minerals		
Calcium	41.0 mg	4%
Copper	0.621 mg	31%
Iron	0.79 mg	0.79%
Magnesium	62.0 mg	16%
Manganese	1.337 mg	67%
Phosphorus	79.0 mg	8%
Potassium	539.0 mg	11%
Sodium	0 mg	0%
Zinc	0.51 mg	3%

GERANIUM FAMILY (GERANIACEAE)

Worldwide, there are six genera and about 750 species in this family.

The flower of filaree RICK ADAMS

FILAREE
Erodium cicutarium

There are about 74 species of *Erodium* worldwide; only *E. cicutarium* is recorded in our area.

Use: The leaves are eaten raw, cooked, or juiced

Range: Prefers lawns, fields, cultivated and disturbed soils, and the fringes of the wilderness

Similarity to toxic species: Since filaree superficially resembles a fern, and perhaps a member of the Carrot or Parsley family (when not in flower), make sure you are thoroughly familiar with filaree before eating any

Best time: Spring

Status: Somewhat common

Tools needed: None

Properties

Filaree is a very common urban weed found in gardens, grasslands, and fields. This annual plant grows as a low-growing rosette of pinnately compound leaves, which are covered with short hairs. The stalk is fleshy. Sometimes, people will think they are looking at a fern when they see filaree. The small, five-petaled flowers of spring are purple, followed by the very characteristic needlelike fruits.

Uses

Filaree leaves and stalks can be picked when young and enjoyed in salads. The leaves are a little fibrous but sweet. I pick the entire leaf, including the long stem, for salads or other dishes. They are best chopped up before being added to salads or cooked dishes such as soups or stews.

You might also enjoy simply picking the tender stems and chewing on them. They are sweet and tasty, somewhat reminiscent of celery. In fact, sometimes, in a dry year, I find that the stem is the only part that I will eat. The leafy section is drier and more fibrous and lends itself better to being added to a stew.

Note the linear seed capsules of filaree.

An overall view of the filaree plant in the field

In wet seasons, the spring growth of filaree is more succulent and tastier. In dry years, the season will be short, and the leaves and stems of filaree will be less desirable.

The filaree plant

RECIPE

Filaree-Up My Cup

If you have a wheatgrass juicer, you can process some filaree leaves and then enjoy the sweet green juice without the fiber.

GOOSEBERRY FAMILY (GROSSULARIACEAE)

This family includes only the *Ribes* genus. There are 120 species worldwide, and some are found all over the United States.

At least four *Ribes* species are recorded in our area; two are gooseberries, and two are currants.

A view of the ripe fruit KYLE CHAMBERLAIN

CURRANTS AND GOOSEBERRIES
Ribes spp.

Use: The fruits are eaten raw, dried, or cooked/processed into juice, jam, and jelly

Range: Found in hilly areas, plains, fields, along rivers, etc.

Similarity to toxic species: When seeing currants for the first time, some folks think they're looking at poison ivy—they've heard the saying "Leaflets three, let it be." But the currant has three lobes per leaf, not three distinct leaflets as does poison ivy.

Best time: The fruits are available in mid-spring

Status: Common locally

Tools needed: None

Properties

Currants and gooseberries are both the same genus, and so we'll treat them together. Both are low shrubs, mostly long vining shoots that rise from the base. The gooseberries have thorns on the stalks and fruits, and the currants do not.

The leaves look like little three- to five-fingered mittens. The fruits of both currants and gooseberries hang from the stalks, with the withered flower usually still adhering to the end of the fruit.

A view of the leaf and ripe fruits

You will find currants or gooseberries widely throughout the area.

Uses

Currants and gooseberries are great fruits, either eaten raw as a snack or dried or cooked into various recipes.

Gooseberries are a bit more work to eat since they're covered with tiny spines. I have mashed them and then strained the pulp through a sieve or fine colander. Then I use it as a jelly for pancakes.

Currants require no preparation, so they can be picked off the stalks and eaten fresh. But make sure they are ripe—they'll be a bit tart otherwise.

The leaves from an emerging stalk of currant

In the old days, the currant was a valuable fruit, dried and powdered and added to dried meats as a sugar preservative. Today, you can just dry the fruits into simple trail snacks. Or you can collect a lot and make jams or jellies or even delicious drinks. And though the currant leaf is not usually regarded as an important food source, some can be eaten in salads or cooked dishes for a bit of vitamin C. They are a bit tough as they get older.

A view of the ripening fruit in the field HELEN W. NYERGES

Rule of thumb on the *Ribes* fruits: If it tastes good, it's good to eat. None are poisonous, but not all are palatable. Remember, some of the forest species are generally not terribly palatable, and some are without significant pulp.

Other Uses
The straight shoots of currants and gooseberries make excellent arrow shafts. Cut the straightest ones green and let them dry.

Cautions
Be sure you've identified currant or gooseberry and that you can tell the difference between these and poison ivy.

Ripe currants

WITCH HAZEL FAMILY (HAMAMELIDACEAE)

Flowers of Hamamelis virginiana

WITCH HAZEL
Hamamelis virginiana

There are at least four known members of the genus *Hamamelis*, with three species in North America.

Use: The plant has long been used as a tonic for rashes, itchy skin, and other skin irritations

Range: Widespread in the mountainous regions

Similarity to toxic species: The flowers are very distinctive, though when not in flower, it's certainly possible to misidentify this plant if you're unfamiliar with it

Best time: Twigs and leaves can be gathered year-round, but the spring is best for medicinal purposes

Status: Somewhat common locally

Tool needed: Possibly clippers

Properties

The yellow, strap-like flowers of this native shrub are among the last blooms to appear in fall, but they are often hidden by the leaves. There are four very slender strap-shaped petals, a half-inch to a quarter-inch long, appearing in mid to late fall. They are pale to dark yellow, orange, or red. In some cases, the conspicuous flowers are the only sign of color in woods when the fall leaves are everywhere on the forest floor.

Robin Wall Kimmerer has a beautiful description of witch hazel in her book *Braiding Sweetgrass*. In describing a walk in the woods in November, she writes, "The flowers are a ragged affair. . . long petals, each like a scrap of fading yellow cloth that snagged on the branch, torn strips that wave in the breeze. But, oh, they are welcome, a spot of color when months of gray lie ahead."

Common witch hazel is a large deciduous shrub, or a small tree, with a picturesque, irregular branching habit that naturally grows along woodland edges. They typically grow from 10 to 25 feet tall but have been observed up to about 40 feet tall. The rounded, dark green leaves are alternately arranged, oval, about two to six inches long and one to four inches wide, with a smooth or wavy margin. The leaves often hang on to the winter branches. The fruit capsules mature a year after flowering, splitting open to expel seeds that are attractive to birds.

The genus name, *Hamamelis*, means "together with fruit," referring to the simultaneous occurrence of flowers with the maturing fruit from the previous year.

A view of the top of a witch hazel leaf MALCOLM MCNEIL

Uses

The main phytochemicals in the witch hazel leaves are polyphenols, including 3 to 10 percent tannins, flavonoids, and up to 0.5 percent essential oil, while the bark has a higher tannin content. Hamamelis water, also called white hazel or witch

The bottom of a witch hazel leaf MALCOLM MCNEIL

hazel water, is prepared from a steam-distillation process using leaves, bark, or twigs and is a clear, colorless liquid containing 13 to 15 percent ethanol having the odor of the essential oil but with no tannins present. Essential oil components, such as carvacrol and eugenol, may be present.

When you purchase a bottle of "witch hazel" at a pharmacy, you're buying the distilled liquid from the cooked twigs. To that distilled product, about 15 percent alcohol is typically added. If you have your own still, you can make your own distilled witch hazel from scratch, starting with the twigs cooked in water.

The witch hazel plant in the woods MALCOLM MCNEIL

According to naturalist Malcolm McNeil, "The twigs and leaves can be gathered throughout the growing season, but they are most effective for medicine in the springtime when the sap is running and the leaves are new and fresh."

As an ingredient and topical agent, witch hazel water is regulated in the United States as an over-the-counter drug for external use only to soothe minor skin irritations.

Witch hazel has long appeared in the US Pharmacopeia for such ail-

Flowers of Hamamelis virginiana

ments as bruises, inflammation, insect bites, burns, hemorrhoids, and so on. The Onondaga and other peoples of the Northeast area heated the witch hazel leaves and branches in water and then used the cooked leaves as a poultice for wounds and bruises. An infusion was also made with the leaves and drunk for diarrhea and dysentery; it was probably effective because of the tannin content.

The name "witch" in witch hazel has its origins in the Middle English *wiche*, from the Old English *wice*, meaning "pliant" or "bendable," and is not related to the word "witch," meaning a practitioner of wicca. The use of the twigs as dowsing rods might have influenced the witch part of the name, according to some sources.

WALNUT FAMILY (JUGLANDACEAE)

The Walnut family contains nine genera and about 60 species worldwide.

A view of the walnut leaf and unripe fruit

BLACK WALNUT
Juglans spp.

There are 21 recorded species of Juglans worldwide, and at least two of those species are found in this area, *J. nigra* and *J. cinerea*.

Use: The nutmeat is eaten. The green walnuts were traditionally used as a fish stunner and the black hulls used for a dye.

Range: Scattered in lower elevation canyons, valley farmland, and urban areas

Similarity to toxic species: None

Best time: Nuts mature in mid- to late summer

Status: Scattered locally

Tools needed: Gloves suggested

Properties

In the greater DC area, there are at least two species of Juglans, not including the English walnut, which can also be found.

Some mature black walnuts in the tree

The black walnut is found in canyons, valleys, and hillsides. In addition, you might encounter the English walnut (*J. regia*), either planted in yards or surviving around old farms and cabins.

This is a full-bodied deciduous tree with pinnately divided leaves. There are typically 11 to 19 leaflets per leaf.

You know what the English walnut that you buy in the store looks like; this one is similar, but there are some important differences. First, all the black walnuts are smaller. They have a soft green outer layer, which turns black as it matures and has long been used as a dye. The shell of the English walnut is thin and easy to crack, but approximately one half of the black walnut is shell. The nut requires a rock or a hammer to crack. The meat in the black walnut is oily and delicious, though there's not as much meat as you'll find in the cultivated English walnut.

Uses

Yes, these are walnuts! But unlike the more commonly cultivated English walnut, these black walnuts are more like hickory nuts, though a bit larger. The shells are hard and thick.

Note that the black walnuts are covered in a fleshy material, which dries when the walnuts are mature and fall. Still, this outer black covering is an excellent dye or pigment material for arts and crafts, but you want to consider wearing gloves when collecting. I once used this to paint children's faces at a day camp, and since the dye takes about two weeks to wash off, I heard from several unhappy parents.

Once you crack open the walnut, you can pick out the edible meat and eat it as is or add it to bread products, cookies, cakes, and even stews and meat dishes. It is a very tasty, oil-rich food and quite a delicacy, but it just takes a lot of work to get to it.

Immature walnuts are sometimes pickled as well. You boil the walnuts and change the water a few times. Then you pack them into jars with vinegar and pickling spices. For details on how to do this properly, read *The New Wildcrafted Cuisine* by Pascal Baudar.

Immature green walnuts were one of the substances used in the old days to capture fish. Indigenous people would crush the green walnuts and toss them into pools of water or the edges of slow-moving streams, and the fish would float to the top. The fish would be scooped out with nets, and then everyone would have dinner!

Note the thick shells of the black walnut. There is typically more shell than meat.

MINT FAMILY (LAMIACEAE)

The Mint family has about 230 genera and about 7,200 species worldwide. Many are used for food, seasoning, and medicine.

A view of the fruits of beautyberry

BEAUTYBERRY
Callicarpa americana

Use: Berries are edible; leaves used as insect repellent
Range: Found somewhat widely
Similarity to toxic species: None
Best time: Fall
Status: Common
Tools needed: Basket for collecting

Properties
This native to the Southeastern United States is a perennial shrub that can reach three to five feet in height. In ideal soil conditions, it can reach up to nine feet tall. It has long, arching branches. The leaves are opposite but sometimes occur in threes. The leaf blade is ovate to elliptic and pointed or blunt at the tip and

tapered to the base. The leaf margins are serrated. The lower surface of young leaves is covered with branched hairs.

Flowers are small, pink, and in dense clusters at the bases of the leaves. The fruit is distinctly colored, rose pink or lavender pink, berry-like, about a quarter-inch long and three-sixteenths of an inch wide, in showy clusters, persisting after the leaves have fallen.

Beautyberry with garden label. Botanists now classify Beautyberry in the Mint family.

The clusters of glossy, iridescent-purple fruit that hug the branches at leaf axils in the fall and winter are this plant's most conspicuous feature.

Beautyberry is found in woods, moist thickets, wet slopes, low rich bottom-lands, and at the edges of swamps. We observed it growing in the garden around the National Museum of the American Indian on the Mall in DC.

Uses

The berries can be eaten raw but are somewhat bland. Jelly is made from the ripe berries. Though the berries are edible raw, they are a bit mealy and even astringent. The flavor and texture are improved quite a bit when a good cook makes these into jelly, jam, and various sauces and juices.

The flavor is described as being reminiscent of elderberry fruit.

Native Americans used root and leaf tea in sweat baths for rheumatism, fevers, and malaria. Root tea is used for dysentery and stomachaches. Root and berry tea is used for colic.

Insect repellent

The American beautyberry has also been used as a remedy to prevent mosquito bites. For some people, rubbing the fresh leaves over the exposed skin has proven effective for repelling mosquitoes for at least a few hours. But for most people, based on anecdotal reports of those who have tried this, rubbing the fresh leaves on the skin provides minimal (if any) protection from mosquitoes. According to a

A view of the leaves and fruits of the beautyberry

USDA study, the chemicals must be distilled from the leaf and then rubbed on the skin.

According to Charles Cantrell, an Agricultural Research Service (ARS) chemist in Oxford, "We actually identified naturally occurring chemicals in the plant [beautyberry] responsible for this activity. Three repellent chemicals were extracted during the 12-month study: callicarpenal, intermedeol, and spathulenol. The research concluded that all three chemicals repulse mosquitoes known to transmit yellow fever and malaria."

MINT
Mentha spp.

A close-up of wild mint

There are 18 species of *Mentha* worldwide, with at least nine recorded in the wild in our area.

Use: As a beverage

Range: Along rivers and wet areas; often cultivated and escaping cultivation

Similarity to toxic species: None

Best time: Mint can be collected at any time

Status: Not common

Tools needed: None

Properties

Mints in the region include spearmint (*M. spicata*), peppermint (*M. piperita*), field mint (*M. arvensis*), and others. In the wild, mints are typically found along streams. They are sprawling, vining plants with squarish stems and finely wrinkled opposite leaves. Crush the leaf for the unmistakable clue to identification. If you have a good sense of smell, you'll detect the obvious minty aroma.

Peppermint and spearmint are usually cultivated in gardens. They sometimes escape cultivation and are found in marshes, ditches, meadows, around lakes, and other moist areas.

The white, pink, or violet flowers of *Mentha* are clustered in tight groups along the stalk, often appearing like balls on the stems. The flowers, though five-petaled, consist of an upper two-lobed section and a lower three-lobed section.

Wild mint. Note the opposing leaves and square stem.

Uses

The wild mints are not primarily food but are excellent sources for an infused tea. Put the fresh leaves into a cup or pot, boil some water, and then pour the water over the leaves. Cover the cup, and let it sit awhile. I enjoy the infusion plain, but you might prefer to add honey or lemon or some other flavor.

We've had some campouts where we had very little food and were relying on fishing and foraged food. Even in off seasons in the mountains, we were able to find wild mint and make a refreshing tea. The aroma is invigorating and helps to open the sinuses. The flavor and taste of mint tea seem even more enjoyable when camping. Also, you can just crush some fresh leaves and add them to your canteen while hiking. It makes a great cold trail beverage and requires no sweeteners.

Sometimes, we add fresh leaves to trout while it is cooking. They add a great flavor. If used sparingly, you can mince some fresh leaves and add them to salads for a refreshing minty flavor. Of course, they can be diced and added to various dessert dishes, like ice cream, sherbet, and so on. Or you can try adding a few sprigs of mint to your soups and stews to liven up the flavor. And if you really want to try something special for your doomsday parties, add a little fresh mint to your favorite pouch of MRE.

If you have a mildly upset stomach, try some mint tea before reaching for some fizzy pill. Mint tea has a well-deserved reputation for calming an upset stomach. The Ojibwa people used the tea to help reduce fevers.

Wild mint JEFF MARTIN

MALLOW FAMILY (MALVACEAE)

The Mallow family includes 266 genera and about 4,025 species worldwide.

According to Dr. Leonid Enari, the Mallow family is a safe family for wild-food experimentation. He cautions, however, that some species may be too fibrous to be easily palatable.

A view of the mallow plant

MALLOW
Malva neglecta

There are 30 to 40 species of *Malva* worldwide. *Malva* is widespread in North America, with six species found in our area.

Use: Leaves raw, cooked, or dried (for tea); "cheeses" eaten raw or cooked; seeds cooked and eaten like rice

Range: Urban areas such as fields, disturbed soils, and gardens
Similarity to toxic species: None
Best time: Spring
Status: Common and widespread
Tools needed: None

Properties

The plants resemble geraniums with their rounded leaves. Each leaf's margin is finely toothed, and there is a cleft to the middle of the leaf to which the long stem is attached. If you look closely, you'll see a red spot where the stem meets the leaf.

A view of the mallow stem and seeds

The flowers are small but attractive, composed of five petals, generally colored white to blue, though some could be lilac or pink. The flowers are followed by the round, flat fruits, which gave rise to the plant's other name, "cheeseweed."

These plants are indeed widespread, mostly in urban terrain and on the fringes.

Uses

When you take a raw leaf and chew on it, you will find it becomes a bit mucilaginous. For this reason, it is used to soothe a sore throat. In Mexico, you can find the dried leaf under the Spanish name *malva* at herb stores, sold as a medicine.

Young mallow seed

Though the entire plant is edible, the stalks and leaf stems tend to be a bit fibrous, so I just use the leaf and discard the stem. These are good added to salads, though they are a bit tough for the only salad ingredient.

The mallow leaf is also good in cooked dishes—soups, stews, or finely chopped for omelets and stir-fries. I have even seen some attempts to use larger

Urban gardener Julie Balaa examines a mallow plant in flower.

mallow leaves as a substitute for grape leaves in dolmas, which is cooked rice wrapped in a grape leaf. I thought it worked out pretty well.

As this plant flowers and matures, the flat and round seed clusters appear. When still green, these make a good nibble. The green "cheeses" (as they are commonly called) can be added raw to salads, cooked in soups, or even pickled into capers. Once the plant is fully mature and the leaves are drying up, you can collect the now-mature cheeses. The round clusters will break up into individual seeds, which you can winnow and then cook like rice. Though the cooked seeds are a bit bland, they are reminiscent of rice. Because mallow is so very common, it would not be hard to prepare a dish of the mallow seed. To really improve the flavor, try mixing the mallow seeds with quinoa, buckwheat groats, or couscous.

The root of the related marsh mallow (*Althaea officinalis*) was once the source for making marshmallows, which are now just another junk food. Originally, the roots were boiled until the water was gelatinous. The water would be whipped to thicken it and then sweetened. You'd then have a spoonful to treat a cough or sore throat. Yes, you can use the common mallow's roots to try this, though it doesn't get quite as thick as the original.

MULBERRY FAMILY (MORACEAE)

Mulberry is a member of the Mulberry family (Moraceae), which includes 37 genera and 1100 species worldwide.

Fruiting mulberry

MULBERRY
Morus spp.

The genus *Morus* includes about 12 species worldwide. In the greater DC area, the genus *Morus* is represented by *Morus alba*, the white mulberry, most commonly found in the wild, and *Morus rubra*.

Uses: Edible fruit

Range: In the wild, you'll find the mulberry in disturbed soils, edges of streams, moist areas, and cultivated areas.

Similarity to toxic species: None

Best time: Summer

Status: Somewhat common in its ideal site

Tools needed: Basket or box. Fruits are fragile, so don't use a bag.

Properties

This appears as a tree with alternate leaves that are unlobed or three- to five-lobed. Produces a catkin and then a fruit that resembles an elongated blackberry.

The fruit is technically referred to as a fruit of many achenes ("seeds"), within the fleshy calyces. Just think elongated blackberry, and you'll get the picture. The white mulberry fruits are white to pink, and the black mulberries are purple.

Uses

Pick these and use them right away. They are best very fresh.

Leaves from a young mulberry tree growing in Virginia

If you go to a nursery to buy a fruiting mulberry tree, you might not find one. Some years ago, nurserymen started switching to non-fruiting varieties because, according to gardeners and homeowners, "the fruits stain the sidewalks." Yes, that happens!

Where I once lived, there was large old mulberry just outside the back door, and yes, it regularly stained the cement walkway. These were the white mulberries, and the stains weren't permanent. I tried to pick and eat as many as I could, and the neighborhood squirrels usually beat me to the fruit on the path.

Mulberries have long been planted as a landscape and park tree, though they have fallen out of favor because of the staining of sidewalks. Sometimes, the

A view of the white mulberries DR. AMADEJ TRNKOCZY

fruiting varieties are planted on large properties so that the fruit attracts birds.

The fruits of any variety can be collected when fresh and eaten as is. They can be dried, too, but the fruit always seems fragile and is best eaten right away.

Of course, jams, jellies, and preserves can be made with the mulberries so you can have some later in the year.

A view of the leaf and fruit

Could the mulberry be regarded as a "clothing tree"? Recall, this is the only thing that the silk moths will eat. I recall a grammar school experiment where we got silk moths from somewhere and fed them mulberry leaves in our classroom. We actually picked the leaves from the schoolyard. Everyone talked about how we'd unroll some of the silk and make a scarf or shirt, but we never got beyond just feeding the larvae.

Other Uses

Archers consider the long, straight branches of the mulberry tree one of the ideal woods for carving a bow.

A mulberry growing in Virginia VICKIE SHUFER

EVENING PRIMROSE FAMILY (ONAGRACEAE)

The Evening Primrose family has 22 genera and about 657 species worldwide.

Evening primrose flower

EVENING PRIMROSE
Oenothera spp.

There are 145 species of *Oenothera* recorded in North America, and nine of these are recorded in this area.

Uses: Roots for food, leaves for food and medicine

Range: Fields, dry land, gardens, farms

Similarity to toxic species: The young leaves can resemble young foxglove leaves. Make sure you know the difference before collecting.

Best time: Spring

Status: Somewhat common

Tools needed: Trowel or digging stick

Properties

Different species of *Oenothera* can be annual, biennial, or perennials. *O. biennis* and several others can be found here in vacant lots, open fields, along riverbanks, and in well-drained soils up to about the 2,000-foot level.

It is an erect biennial with alternate leaves. When the plant first emerges, you will observe the rosettes of elliptical leaves. The lower leaves are stalked, whereas the upper leaves, as the plants get taller, are sessile or stalk-less.

The fragrant flower is composed of four yellow petals, four sepals, and eight stamens. Each stamen supports a four-lobed stigma, which is the pollen-bearing part of the stamen. A narrow fruit follows the flowers, which is a four-celled capsule about two inches long.

A view of the leaf of O. biennis

Uses

The roots of the first year are generally sought for food. Dig the youngest roots, and then peel, and cook them by boiling or by cooking in an underground pit. Younger roots are best, as the older ones get tough and strongly flavored. Sometimes, a change of water will be required to make the roots palatable.

Slice the roots, and sauté or cook in stews. (Boil them first if the roots are tough.) Serve with butter or cheese.

Is this a really good food, or is it marginal? I think it depends on how hungry you are and the age of the

The overall size of the flowering evening primrose plant

root. Older roots are simply not as tender or tasty as the young ones.

We've tried some leaves too—both raw and cooked—and they get mixed reviews. The young ones are OK when used sparingly in salads, but you're most likely to enjoy them a bit better when steamed or boiled and blended with other foods.

The astringency of the leaves makes them useful for coughs and sore throats, either made into an infusion or simply chewed.

The flowers of the evening primrose

POKEWEED FAMILY (PHYTOLACCACEAE)
POKEWEED
Phytolacca americana

The mature fruit of poke

There are 35 species of Phytolacca worldwide, with at least three species in our area.

Use: Young shoots for cooked greens; fruits can be cooked and eaten

Range: Prefers disturbed soils, fields, lawns

Similarity to toxic species: This *is* a "toxic" species—so make sure you properly process it if you intend to eat it!

Best time: Spring for the young shoots

Status: Fairly common

Tools needed: Collecting bag, knife

Properties

Pokeweed is a bushy, herbaceous perennial that can grow up to seven feet and taller in some cases. It has a large fleshy taproot that can grow up to four inches in diameter. Several stems can emerge from the taproot in the spring. The stems become deep red to purple and become branched in the upper part of the plant as it grows older. The large leaves are alternately arranged and are smooth, shiny, and fleshy. In the very young stages, there are several other plants that it could be confused with, some of which can be more toxic than the poke.

The individual leaf shape is elliptical to egg shaped and tapering to a point at the tip. The leaf stalk can be from a half-inch to two inches long. It is not unusual to find mature poke leaves that are a foot long.

The whitish-green flowers are produced in unbranched nodding clusters around six inches long, at the ends of the stems and upper branches. Flowers are approximately a quarter-inch wide, composed of five sepals that resemble petals. The dark purple berries follow, formed in clusters in the fall. These berries are eight- to 10-chambered, shiny, and with one seed per chamber. Berries are round, flattened, and turn from green to dark purple as they mature. The clusters of the fruit become heavy as they mature, and the clusters droop on the plant. The plant readily reproduces from the seeds of the fruit when they drop or are spread by animals.

The young poke shoot MALCOLM MCNEIL

Uses

Poke is fairly common and widespread in our area and throughout the eastern part of the United States. It's found growing in both the urban areas and rural areas, as well as in wilderness sites.

I first encountered pokeweed while living on my grandfather's farm during the summers. It's a fairly distinctive plant, so once you've seen it a few times, you won't confuse it with anything else.

The root of the poke plant MALCOLM MCNEIL

A big misconception about eating poke resulted from the popular 1968 song by Tony Joe White called "Polk Salad Annie." The song was about a girl who picked pokeweed and knew how to prepare it. But if you're from the South, you knew that White should have spelled it "Poke Sallet," since "salad" suggests you eat poke raw—you can't!—and "sallet" refers to cooked greens, which is the proper way you must prepare poke.

So, the first poke shoots are collected, boiled, and the water discarded. The greens are then boiled and the water discarded again. Then, after all that, you can season and eat the greens with any of the traditional dressings, like bacon fat or butter, or add the greens to soup or casseroles.

A view of the red stalk of mature poke. Found in suburban Alexandria

Here is how herbalist Susun Weed describes the processing of poke greens: "To make your own sallet: Collect very young poke greens as early as possible in the season (late April to mid-May in the Catskills; as early as February in Georgia). Pour boiling water over the greens and boil them one minute. Discard water. Add more boiling water and again

Roman inspects the maturing poke plant.

boil the greens for one minute. Discard the water. Do this at least twice more before attempting to eat the greens. If you fail to leach out the poisonous compounds—or are foolish enough to attempt to eat poke leaves raw—your mouth and throat will feel like they are on fire, you may vomit, and you will no doubt have copious diarrhea."

There seems to be scant data in the literature about the use of pokeberries for food. Mostly, I read that the berries are poisonous and could kill you and that you should never eat them. According to Donald R. Kirk, author of *Wild Edible Plants of Western North America*, "The berries are reported by some to be edible and by others to be poisonous. Those that use them make them into pies and jellies."

On our farm, we cooked the berries and made pie. My uncle grew up eating poke pie that his father made and always enjoyed it. As a new poke eater in my late teens, I simply boiled the mashed fruits, strained out the seeds, sweetened it, and added it to a pie crust. It did have a characteristic and unique flavor, but we all ate it, and no one got sick. Since everyone's body chemistry is different, I would suggest that you proceed with caution if you decide to try this.

The author photographing a poke plant in a residential area of Virginia HELEN W. NYERGES

Susun Weed points out that the small seeds in the berries are poisonous. "Lucky for us," she writes in sagewoman.com, "they are too hard for our teeth to break open. I have had pokeberry jam (no worse than blackberry jam, that is, seedy) and pokeberry jelly (ah, no seeds) and pokeberry pie (seedy). Since children are attracted to poke plants and since the berries leave telltale stains on children's mouths and since many parents are frightened if their child eats anything wild and since medical personnel know little about poke except that it is poisonous, lots of kids have their stomach pumped (for no good reason, since they can't break open the seeds either) after investigating the taste of poke berries."

These days, when I read about people getting sick from products made from pokeberries, I tend to be a bit more cautious. I would suggest mashing the fruits and removing all seeds and then boiling them thoroughly. Yes, there are those who always bring up the fact that hard-boiling can destroy some vitamins. Yes, but it's either that or vomiting! I prefer the former! Or just avoid these berries altogether and stick to strawberries and grapes.

The berries can also be crushed and made into a paint for artworks and possibly fabrics. I experimented a little with the ink and found it to be a refreshing color when brushed onto white art paper.

Caution:

Yes, you might read in some herbal literature that the root is good for certain maladies. I suggest that you just leave the root in the ground! The root has sometimes been suggested as a remedy for various inflammations, mastitis, rheumatism, and so on by trained herbalists.

Nevertheless, the root should be considered toxic unless you have taken the time to learn how to use it properly. Consider the following case: On January 26, 1976, Beverly Dixon was strolling through a Madison, Wisconsin, shopping mall, and largely on impulse, she purchased a box of "dried powder of pokeweed." It came with instructions on how to brew it into tea, and it had an attractive picture on the box of the beautiful berries.

An hour and a half after Dixon drank the tea, she began to vomit and had severe cramps and pain. Her husband rushed her to the intensive care unit at the University Hospital, where her stomach was pumped. She was hospitalized 22 hours in all and, fortunately, recovered.

Remember, poke *is* a poisonous plant, and so if you eat the young shoots, be sure to cook them twice as described here so you won't get sick.

The poke plant growing on the grounds of the Museum of the American Indian on the Mall

PLANTAIN FAMILY (PLANTAGINACEAE)

The Plantain family has 110 genera and approximately 2,000 species worldwide.

Broadleaf plantain going to seed RICK ADAMS

PLANTAIN
Plantago major and *P. lanceolata*

There are about 250 species of *Plantago* worldwide, with at least seven found in our area.

Use: Young leaves used for food; seeds used for food and medicine

Range: Prefers lawns, fields, and wet areas

Similarity to toxic species: None

Best time: Spring for the leaves; late summer for the seeds

Status: Fairly common

Tools needed: None

Properties
Plantain is as common an urban weed as dandelion, though not as widely known. It's usually found in lawns and fields but also in wet areas. We observed lots of plantain growing on the south lawn of the White House.

A patch of the broadleaf plantain

Broadleaf plantain

All the leaves radiate from the base in a rosette fashion, with the basal leaves typically from about two to six inches in length. *P. lanceolata's* leaves are narrow, prominently ribbed with parallel veins. *P. major* has broad, glabrous leaves up to six inches long, roundish or ovate shaped. Both have leaves that are covered with soft short hairs. Both originated in Europe.

The flowers are formed in spikes (somewhat resembling a miniature cattail flower spike), usually just a few inches long, and on stems that are typically no more than a foot tall. Each greenish flower is composed of four sepals, a small corolla, and four stamens (sometimes two). The flowers are covered by dry, scarious bracts. When the spikes are dry, you can strip off the seeds and winnow them.

Roman examines an exceptionally large broadleaf plantain leaf.

Uses

The young tender leaves of spring are the best to eat; use in salads or as you would spinach. The leaves that have become more fibrous with age need longer cooking, and they are best finely chopped or pureed and cooked in a cream sauce. The leaves have a mild laxative effect.

The seeds can be eaten once cleaned by winnowing. They can be ground into flour and used as you would regular flour or soaked in water (to soften) and then cooked like rice. Once cooked, the seeds are slightly mucilaginous and bland. They can be eaten plain or flavored with honey, butter, or other seasonings.

Cooked plantain leaves have been used as a direct poultice on boils. Plantain is a vulnerary (promotes healing) and is noted for its styptic, antiseptic, and astringent qualities. Native people used the cooked leaves as a poultice for wounds.

Plantain leaf, crushed or chopped and used as a poultice, is perhaps the best herb to use for puncture wounds to the body (knife wound, stepping on a nail, etc.).

Early American colonists applied crushed plantain leaves on insect and venomous reptile bites. I've applied the crushed leaves to mosquito bites on many occasions and found only minimal relief. I obtained greater relief by doing the same with chickweed leaves.

Angelo Cervera examines the maturing narrowleaf plantain.

Narrowleaf plantain

VERONICA, AKA SPEEDWELL
Veronica americana

The asymmetrical flowers of veronica RICK ADAMS

The *Veronica* genus has about 250 species worldwide, at least 15 of which are found in this area.

Use: The entire plant (tender stems and leaves) above the root can be eaten

Range: Grows in slow-moving waters, in the same environment as watercress

Similarity to toxic species: None

Best time: Spring and summer

Status: Somewhat common

Tools needed: None

Properties
Veronica americana is a native and has been confused with watercress because they both grow in water. The resemblance is superficial because there are obvious differences. The veronica has a simple leaf about one or two inches long, whereas the watercress has pinnately divided leaves, very much like many of the members of the Mustard family. The watercress has a typical mustard flower formula with the four petals arranged like a cross, and its color is white. But the veronica flower is lavender and asymmetrical with four petals, the upper one being wider than the others.

Uses

If I have no concerns about the water's safety from which I've picked the veronica, I add it to salads. It is not strongly flavored, and you can use the entire plant. Just pinch it off at water level (no need to uproot the plant), rinse it, and then dice it into your salad. No need to pick off just the leaves—eat the entire above-water plant since the stalk is very tender. Since it's so bland, you can mix it with stronger-flavored greens in your salad. It goes well with watercress, as well as any of the mustards.

Veronica also goes well with soup dishes and stir-fries. It never gets strongly bitter, like watercress, and it never really gets fibrous. It's a mild plant that's fairly widespread in waterways.

If you live near a waterway where veronica grows, you'll find that it's a good plant to use in a variety of dishes where you might otherwise include spinach. Try some gently sautéed with green onions and add some eggs to make an omelet. Try a cream soup into which you've gently cooked some veronica greens.

We've included veronica stems and leaves in nearly every dish we've made when out on our field trips. The leaves are tender and mild and go well with just about anything else. We've cooked them with wild mushrooms, added them to egg dishes, blended them into miso soup and nettle soup, added them to wild salads, and blended the diced veronica with rice dishes.

Veronica rising from a bed of watercress

The veronica plant growing in shallow, slow-moving water

BUCKWHEAT FAMILY (POLYGONACEAE)

The Buckwheat family has 48 genera and about 1,200 species worldwide.

A view of the growth pattern of the sheep sorrel plant

SHEEP SORREL
Rumex acetosella

There are about 190 to 200 species of *Rumex* worldwide, with at least eight species in our area.

Use: The leaves are good raw in salads and can also be added to various cooked dishes

Range: Found in the higher elevations, often around water, and often near disturbed soils and in urban areas

Similarity to toxic species: None

Best time: Spring to early summer

Status: Can be abundant locally

Tools needed: None

Properties

Sheep sorrel is native to Europe and Asia. It is common and widespread and is recognized by its characteristic leaves, which are generally basal, lance to oblong shaped, with the base tapered to hastate or sagittate. In other words, it looks like an elongated arrowhead. When the seed stalk matures, it is brown, reminiscent of the curly dock seed stalk but much smaller.

Uses

Where the plant is common, you can pinch off many of the small leaves to add to salad or even to use as the main salad ingredient. I've enjoyed sheep sorrel salads with just avocado and dressing added. The leaves are mildly sour, making a very tangy salad. However, I generally prefer to add these leaves so that they comprise no more than about a quarter of the salad ingredients. When added to soups or stir-fries, the flavor becomes subdued, though I still use sheep sorrel greens that comprise no more than about a quarter of the total greens for a dish.

The flavor is somewhat similar to the leaves of oxalis, though not as strong. They can be effectively added raw to other foods like tostadas (in place of lettuce) or sandwiches. They add a bit of a tang when added to soups and stews and can be very effective at livening up some MREs.

The shape of the sheep sorrel leaf resembles an arrowhead.

Collecting some of the wild young leaves of sheep sorrel for lunch

RECIPE

Shiyo's Garden Salad

Rinse a bowl full of young sheep sorrel leaves. Add at least one ripe avocado and one ripe tomato, both diced. Toss with some Dr. Bronner's oil and vinegar dressing. Eat it outside where the wind can blow your hair.

CURLY DOCK
Rumex crispus

A view of dock leaves. These are not too young and not too old. Some can still be used in salads, and most can be used for soup and other cooked dishes. RICK ADAMS

There are about 190 to 200 species of *Rumex* worldwide, with at least eight in our area.

Use: Young dock leaves eaten raw or cooked; seeds harvested and added to various flours; stems used like rhubarb

Range: Prefers wet areas but can be found in most environments

Similarity to toxic species: None

Best time: The leaves are best gathered when young in the spring. Seeds mature in late August, and they may be available for months

Status: Common and widespread

Tools needed: None

Properties

Curly dock is a widespread perennial plant. It is originally from Europe, and today it is found widely in our area and worldwide. Though it has many good uses, it is often despised and poisoned because it not only survives well but often takes over entire areas.

The root looks like a dark orange carrot, and the spring leaves grow directly from the root. The young leaves are long and linear and are wavy or curved on their margins. Those curved margins are the origin of the name "curly." The leaves can be over a foot long and pointed.

As the season progresses, the flower stalk rises, and it can reach about four feet, even taller in ideal conditions. The seeds are formed with three to each unit, with a papery sheath around the seed. They are green at first and then mature to a beautiful chocolate brown.

Some of the other species of *Rumex*—such as *R. obtusifolia*—might be confused for this one, though they are all used in pretty much the same manner.

A view of the immature green dock seeds

Uses

You can make meals from both the leaves and the mature seeds of curly dock. Let's start with the leaves.

Pick only the very youngest leaves for salad, the smaller ones before the plant has begun to send up its seed stalk. These will not be too tough, and the flavor will be sour, somewhat like the French sorrel. You can just rinse them, dice them, and add them to salads. I've had *only* these for salad, with dressing and avocado, and it was good, but only because the leaves were young.

Older leaves are best boiled like spinach, or—ideally with the midrib removed—sautéed with potatoes and onions. Or you can just add some to soup and stews. The leaves change color and darken a bit upon cooking, and the cooking softens up the tougher older leaves. But you really want to cook older leaves, as they are tougher and bitter and astringent, all of which is reduced somewhat by cooking.

I have seen the brown seed spikes sold in floral supply shops as "fall decoration," and they are very attractive. Those little seeds can be stripped off the stalks with your hands and then

Sauteed dock leaves, hardboiled eggs, tomatoes, and red onions make an excellent blend.

rubbed between the hands to remove the wing from the seed. You don't have to be too picky here, as it can all be used. I blow off the wings and then mix the seed fifty-fifty with flour for pancakes and sometimes bread. You could also toss some seeds into soup to increase the protein content.

I've seen some folks go to the trouble of winnowing and then further grinding the seeds in a mill to get a fine flour. I don't usually bother to do this, but it does produce a finer flour, which is a bit more versatile than the whole seeds. For example, a fine flour can be mixed fifty-fifty with wheat, blended, and put through a pasta machine to make a curly dock seed pasta, which tastes really good.

The leaf stems are tart and sour but often make a good nibble. Young stems can be processed and used like rhubarb for pies.

The maturing seed stalks of curly dock

PURSLANE FAMILY (PORTULACACEAE)

The Purslane family has recently been redefined by botanists as having only the one genus, with about 100 species worldwide, and only this species is found here. Many of the plants that were formerly in this family are now a part of the Miner's Lettuce family. This family is considered by botanist Dr. Leonid Enari to be entirely safe for consumption.

A patch of purslane

PURSLANE
Portulaca oleracea

Use: Entire aboveground plant can be eaten raw, cooked, pickled, and so on

Range: Prefers disturbed soils of gardens and rose beds; also found in the sandy areas around rivers

Similarity to toxic species: Somewhat resembles prostrate spurge. However, spurge lacks the succulence of purslane. Also, when you break the stem of spurge, a white milky sap appears, which does not appear with purslane.

Best time: Spring into summer

Status: Relatively common

Tools needed: None

Properties

Purslane starts appearing a bit later than most of the spring greens, typically by June or July. It is a very common annual in rose beds and gardens, though I do see it in the wild occasionally, typically in the sandy bottoms around streams.

The stems are succulent, red-colored, and round in the cross section. The stems sprawl outward from the roots, rosette-like, just lying on the ground. The leaves are paddle shaped. The little yellow flower is five-petaled.

The sprawling purslane plant

Uses

When you chew on a fresh stem or leaf of purslane, you'll find it mildly sour and a bit crunchy. It's really a great snack, though I like it a lot in salads. Just rinse to get all the dirt off, dice, add some dressing, and serve. Yes, add tomatoes and avocado if you have any.

Add it to sandwiches, tostadas, even on the edges of your chiles rellenos and huevos rancheros. I've also eaten it fried, boiled, baked (in egg dishes), and

A bundle of young purslane for sale at a farmers market

Purslane growing outside an Alexandria bakery

probably other ways too. It's versatile, tasty, and crisp. It really goes with any-thing, and it's very nutritious.

If you take the thick stems, clean off the leaves, and cut them into sections of about four inches, you can make purslane pickles. There are many ways to make pickles; my way is to simply fill a jar with purslane stems, add raw apple cider vinegar, and let it sit for a few weeks. (I refrigerate it.)

According to researchers, purslane is one of the richest plant sources of omega-3 fatty acids. That means not only is it good, but it's also good for you!

Henry David Thoreau, who lived in Massachusetts, was a man who liked to think about the meaning of life. During one introspective period in his life, he built his 10-by-15-foot cabin on his friend's property (Ralph Waldo Emerson) and lived very simply, growing food, fishing, spending time talking with the local Indians, and foraging for his meals. He wrote about this time of his life in his book *Walden*. He wrote, "I have made a satisfactory dinner off a dish of purslane which I gathered and boiled. Yet men have come to such a pass that they frequently starve, not from want of necessaries, but for want of luxuries." I admired Thoreau, and when we studied him in high school, my teacher called him a "philosopher." Years later, I met some of the descendants of peers of Thoreau, and they told me that Thoreau was generally thought of by his peers as more of a bum than a philosopher. I realized that people working at regular jobs in town would have that perspective, despite the great value of Thoreau's many contributions to our culture.

Purslane Salsa

2 cups chopped tomatoes

2 ½ cups chopped foraged purslane

¾ cup chopped onions

3 garlic cloves

1 cup apple cider vinegar

¼ cup sugar

1 large Oregon myrtle leaf

½ cup chopped cilantro and some herbs from the garden (thyme, etc.)

Salt and pepper to taste

Place all the ingredients except the cilantro and herbs in a pot, bring to a boil, and then simmer until the right consistency (light or chunky). Add the cilantro and herbs at the end, and salt and pepper to taste.

Pour into jars, close the lids, and place in the fridge. It should be good for at least a month.

—RECIPE FROM PASCAL BAUDAR

Note the reddish stem and paddle-shaped leaves of purslane.

ROSE FAMILY (ROSACEAE)

The Rose family contains 110 genera and 3,000 species worldwide.

Serviceberry leaf and fruit JOHN DOYEN

SERVICEBERRY
Amelanchier alnifolia and six other spp. of *Amelanchier*

The *Amelanchier* genus consists of about 10 species, which are found all over North America. At least seven are found in our area.

Use: Edible berries

Range: Most common in riparian and moist hillside areas

Similarity to toxic species: None

Best time: Late summer and fall

Status: Somewhat common

Tools needed: Collecting basket

Properties
Serviceberry is a large shrub or small tree with deciduous leaves, often forming in dense thickets.

The twigs of this native are glabrous, and the leaf is elliptical to round, with obvious serrations, generally serrated above the middle of the leaf. The flowers

A view of the ripe fruit LOUIS-M. LANDRY

are five-petaled, white, fragrant, and in clusters of a few to many. The fruit is a pome, berrylike, generally spherical, bluish-black to purple in color, with a waxy outer skin. Each fruit contains two seeds. The shape somewhat resembles a tiny pomegranate.

Uses

The ripe fruits are good to eat raw, dried, or prepared into jams. Fruits of several species of *Amelanchier* were used for food by various Native American tribes, and all members of this genus are edible. Fruits ripen in late spring and into the summer.

Native peoples ate these fruits fresh, or they dried them for later use. The ripe berries were mashed with water into a paste by the Atsugewi and then eaten fresh. Several of the Western tribes were known to dry these fruits and then shape them into loaves for future use.

The berries would remain sweet when dried and could be reconstituted later when added to water. In some cases, this would be served as a sweet soup. With sugar and flour added, these fruits have been made into a pudding. The fruit can be dried, ground, and used in a pemmican mix.

HAWTHORN
Crataegus spp.

The ripe hawthorn fruit KYLE CHAMBERLAIN

There are about 300 species of Crataegus worldwide. About 16 are found in the greater Virginia area, such as the Cockspur hawthorn.

Uses: Fruit edible

Range: Widespread, but prefers riparian thickets

Similarity to toxic species: No

Best time: Late summer

Status: Common

Tools needed: Basket

Properties

Hawthorn is a deciduous tree, or large shrub, found widely throughout the area.

The leaves are ovate, alternately arranged, with teeth that all point upward to the tip of the leaf.

The flowers are white to pink and are about a quarter-inch across, forming flat terminal clusters. Each urn-shaped flower has five petals and many stamens. When fully in flower, there is a strong flavor, often described as unpleasant. The red to black fruits are in clusters, with each fruit having two to five seeds.

The plant is covered with one-to-three-inch thorns along the branches.

Uses

The ripe berries are pleasant-flavored, though they are only mildly sweet, seedy, and somewhat dry. Well worth consuming fresh when encountered, the abundant hard seeds make processing and storage tricky. The pectin-rich pulp, squeezed through cloth or one's fingers, will make a mild-flavored fruit leather. Alternately, commercial strainers can be used. The fruits have heart-supporting properties and help prepare the metabolism for winter.

The fruits were used fresh but also dried (often mashed into cakes) and stored for future use. Sometimes the fruits were soaked in water and made into a drink.

Among some tribes, these berries were only used sparingly or regarded as "famine food" when nothing more desirable was available.

Fruits were also mixed into pemmican, which was typically a mix of dried and crumbled meats and suet.

Cautions

Consume the fruits with caution and moderation the first time you try them, as some users have reported stomachaches after consuming these.

STRAWBERRY
Fragaria spp.

The strawberry plant in flower JEAN PAWEK

Fragaria contains 20 species worldwide, with two found wild in the area are covered by this book.

Use: Edible berries; leaves used for tea

Range: Can be found on beaches and in farmlands, fields, and meadows

Similarity to toxic species: None

Best time: Spring and summer

Status: Widespread

Tools needed: Collecting basket

FORAGER NOTE: Strawberry leaf tea (made by infusion), though not strongly flavored, is popular in many circles. It is high in vitamin C and generally used as you'd use blackberry leaf or raspberry leaf tea. It's a mild diuretic, has astringent properties, and is regarded as a tonic for the female reproductive system. When made stronger, the tea is said to be good for hay fever.

Properties

If you've grown strawberries in your yard, you will recognize these two wild strawberries.

Ground cover of the wild strawberry plant

The wood strawberry (*F. vesca*) is found in partial shade throughout our area. Receptacle is about five to 10 millimeters; leaf petiole is generally three to 25 centimeters.

The mountain strawberry (*F. virginiana*) is found in the higher elevations in meadows and forest clearings. Receptacle is more or less about 10 millimeters; leaf petiole is generally one to 25 centimeters.

Their leaves are all basal, generally three-lobed, with each leaflet having fine teeth. They look just like the strawberries you grow in your garden but smaller.

Technically, the strawberry berry is an aggregate accessory fruit, meaning that the fleshy part is derived not from the plant's ovaries but from the receptacle that holds the ovaries. In other words, what we call the "fruit" (because, obviously, it looks like a fruit) is the receptacle, and all the little seeds on the outside of the "fruit" are technically referred to as achenes, actually one of the ovaries of the flower, with a seed inside it.

Though the wild strawberry prefers higher elevation forests and clearings, it is found widely throughout the state.

Wild strawberries are pretty easy to identify. When the average person sees one, especially if it's summer and the plant is in fruit, he or she will typically say, "Hey, look, isn't that a wild strawberry?" Strawberries are so widely known that just about everyone recognizes them when they see them, even though the wild varieties are significantly smaller than the huge ones that can be found in the markets. Cultivated strawberries can get to be about two—even up to three—inches long. That's huge! By contrast, a wild strawberry is between a quarter and a half an inch long. A half-inch wild strawberry is a big one!

Though they may be smaller, the wild strawberries are typically sweeter, firmer, and tastier. Yes, it may take longer to collect them, but you'll find that it's worth it.

Uses

You use these in every way that you'd use cultivated strawberries. Eat them as is, dry them, make them into jams and jellies, put on top of ice cream and pancakes, and so on.

When I was in my teens, I began learning how to garden, and I tried growing strawberries in tiers. It was a lot of work to get some fragile fruits, but they were still far superior to any strawberries in the market. But I gave it up after a few seasons since I felt it was too much work for the result.

I was amazed to see wild strawberries literally blanketing the floor of my grandfather's farm, forming a mat as the runners spread in all directions, creating new plants. In the summer, I would go out with a little basket to collect the fruits. But the wild strawberries are small, even tiny. A half-inch-across wild strawberry would be considered "big." So, I collected, and I nibbled, and rarely did I ever bring many back to the kitchen.

These are well worth the effort to pick them, even if you eat them on the spot.

Strawberries were eaten by all the Native people.

The leaves can also be infused into a medicinal tea. Since there is hardly any flavor, it is often mixed with a stronger herb, such as mint.

Fruit of wild strawberry—F. vesca DR. AMADEJ TRNKOCZY

APPLE AND CRABAPPLE
Malus spp.

A view of the wild crabapple tree in fruit WILLIAM J. HARTMAN

Worldwide, there are about 25 species of the *Malus* genus, which includes all of our domestic apples. In our area, we find at least eight species of *Malus*, including *M. pumila* (the domestic apple and its varieties) and the crabapple (*Malus* hybrids).

Use: Edible fruits

Range: Occurring widely throughout North America, from coast to coast, from Canada to Mexico. Found in moist woods, fields, swamps, and open canyons from sea level to moderate elevations in the mountains.

Similarity to toxic species: None

Best time: Fall through winter

Status: Common

Tools needed: Collecting basket

Properties

If you've ever seen a domestic apple tree in an orchard or backyard, you know what the tree looks like. In the wild, these will be small trees, often in thickets.

You'll look at the leaves and the fruit, and you'll say to yourself, "Boy, that sure looks like an apple." Yes, it IS an apple; it is a wild apple.

Each leaf is lanceolate to ovate-lanceolate, four to 10 centimeters long, pointed, serrate, occasionally with a lobe on one or both margins, and deep green above and paler beneath.

The floral inflorescence typically has five to 12 flowers and is flat-topped. There are five white petals and 20 stamens, shorter than the petals; the styles usually come in threes.

These fruits are fleshy, round to obovoid, about 10 to 16 millimeters long. The color can range from yellow to purplish red.

A wild crabapple ZOYA AKULOVA

Uses

Remember, if you know what an apple looks like, you'll recognize the crabapples or crabbies. Probably everyone who has seen one for the first time has picked one, chewed on it, and spit it out because it was too sour. These are not great to eat raw like you'd eat a regular domestic apple, but they are still a great find. The first time I saw a crabapple tree was when I was away from home visiting my cousin to the north in Ohio. We were in an area where the crabapples grew thickly, and I picked one for the first time, recognizing that it was an apple. I bit

A view of wild crabapples WILLIAM J. HARTMAN

into it and thought it was good. "We don't eat those," said my cousin disdainfully. "They give you the runs."

In fact, properly prepared, these can be quite good. I've cooked them and run them through a sieve to get rid of the seeds and skin and made a great crabapple sauce. You can sweeten them with a bit of honey, as you wish.

Wild crabapples JEAN PAWEK.

The fruits can be dried as a snack for later or be mashed and added to baked goods. The fruits can be cooked, mashed and strained, and used as the basis for an apple drink. The cooking mellows the flavor, but you still might want to add some honey, and this goes particularly well with a cinnamon stick.

You can pretty much do anything with crabapples that you'd do with cultivated apples, such as cooking and mashing up a batch, spreading it thin in a pan, and drying it for fruit leather.

If you're good in the kitchen, you can cook up a batch of the small crabapples to make jams or jellies.

The fruits often remain on the trees and will lose a bit of their tartness and even sweeten up a bit if you harvest in the wintertime.

You might be surprised how many gone-feral apple trees you can find that still produce fruit. Some of the best apples I've ever eaten were picked in orchards that had been abandoned at least a decade earlier, yet they still consistently produced quality fruit.

Cautions

The seeds of crabapples, and domestic apples, are toxic because of a small amount of a cyanide compound. But you'd have to eat MANY apples to cause sickness, and if you don't chew the seeds, they will just pass through your body. Death had occurred when an adult male chewed up about a half cup of pure seed at one sitting—he liked the flavor of the individual seeds and didn't realize that a larger volume could be toxic. Fortunately, cooking and drying help to break down this chemical, significantly reducing any danger.

It's possible that you might get "the runs" (i.e., diarrhea) by eating a lot of the raw fruits. "A lot" can vary from person to person, so if you enjoy eating these raw, go slowly at first and monitor what happens to your body.

WILD CHERRIES/STONE FRUITS
Prunus spp.

Prunus virginiana BOB KRUMM

There are about 400 species of *Prunus* worldwide, whose common names generally include cherry, almond, apricot, and plum. At least 19 species of *Prunus* are recorded in our area, such as the chokecherry (*P. virginiana*) and the black cherry (*P. ceratina*).

Use: Flesh of the fruit in jams and jellies; meat of the large seed processed into a flour

Range: Canyons, lowland forests, hillsides, farmland, urban areas

Similarity to toxic species: In a sense, this is a toxic plant. The leaves are mildly toxic—see CAUTION section below. And the seeds are mildly toxic if eaten raw.

Best time: Fruits mature around July into August

Status: Common

Tools needed: Collecting box

Properties

One way to identify the plant is to crush the leaves, wait a few seconds, and then smell them. They will have a distinct aroma of bitter almond extract, which is your clue that the leaf contains "cyanide" (hydrocyanic acid).

The fruits are very much like cultivated cherries, except the color is darker red, almost maroon, sometimes even darker. The flesh layer can be very thin in dry years and thicker in the seasons following a good rain. Like domestic cherries, there is a thin shell and the meaty inside of the seed.

Uses

The fruit of wild cherries makes a great trail nibble. I usually see them in August when they ripen, when the trail is hot and dry, and the fruit makes a refreshing treat, if not too sour. But don't eat too much of the raw fruit, or diarrhea might result.

The wild cherry also has a hint of bitterness. The fruit can be cooked off the seeds, and the pulp can be made into jellies, jams, and preserves. You can also make a fruit leather by laying the pulp on a cookie sheet and drying it.

The fruit of wild cherry—P. virginiana LOUIS-M. LANDRY

In the old days, all Native people enjoyed the flesh of the cherry; some considered the seed to be the more valuable part of the fruit. The seeds were shelled, and the inside meat was cooked to reduce the cyanide. The cooked seeds, once ground into mush or meal, were then used to make a sweet bread product or added (like acorns) to stews as a gravy or thickening agent.

According to naturalist Malcolm McNeil, "Many of the cherry trees you may see in the greater DC area will not be trees that bear fruit. The trees that are well-known for the Cherry Blossom Festival are Japanese-imported species that do not produce edible fruits. If you wish to grow your

A wild cherry (P. virginiana) leaf LOUIS-M. LANDRY

own cherry tree that fruits, try the native Black Cherry tree."

The bark was boiled by Native people and used as a cough and sore throat remedy as well as for treating diarrhea and headaches.

Cautions

If you crush the leaf, it will impart a sweet aroma like the bitter almond extract used in cooking. That's the telltale aroma of cyanide, so don't use the leaf for tea.

Eating a large volume relative to your body size could cause stomach pains or diarrhea, so exercise caution.

Prunus avium is one of the many species of Prunus commonly planted as an ornamental. TOM NYERGES

WILD ROSE
Rosa spp.

Wild rose flowers

There are about 100 species of *Rosa* worldwide, which hybridize freely. At least 13 species can be found in the wild in our area, not including varieties and not including the large array of cultivated roses grown in gardens. All species are widespread at low and middle elevations.

Use: Fruits eaten raw or cooked and made into jam or tea; wood of the straight shoots useful for arrow shafts

Range: Typically riparian but found in many areas; cultivated roses are common in urban areas

Similarity to toxic species: None (but be wary of eating fruit or flower from roses where various commercial fertilizers and insecticides have been used)

Best time: Fruits mature in summer

Status: Common

Tools needed: Clippers, possibly gloves

Properties

Wild roses are more common than most people realize. In the wild, they are more likely found in wet areas, though this is not a hard-and-fast rule. The wild rose flowers are five-petaled, not the multiple-petaled flowers that you find on hybridized roses. After the flowers mature and fade, the fruit develops, often called the "hip," which is usually smaller than a grape. The fruit is bright orange.

The leaves are oddly divided into three, five, or seven petals, and the stalks are covered in thorns. If you've ever had rose bushes in your yard, you have a pretty good idea of what the wild rose looks like.

The wild rose is often in dense thickets. If it gets cut down, or after a burn, there will be many straight shoots in the new growth.

Uses

For food, we use the flower and the fruits. The flowers have long been used to make "rose water" and could also be used to make a mild-flavored infusion. The petals make a flavorful, colorful, and nutritious garnish for soups and salads.

The fruits—commonly called "hips"—are one of the richest sources

The new shoots of the wild rose plant

Ripe fruit ("hip") of the wild rose

Wild rose flower RICK ADAMS

of vitamin C. The fruits can be eaten fresh, but you should first split them open and scrape out the more fibrous insides. They are typically a bit fibrous, with a hint of bitterness. The fruits are more commonly cooked into a tea or made into jellies.

Some "old-school" archers consider the rose shaft one of the finest woods to use for making arrows, assuming you cut the new straight shoots. You need to then ream the shaft through a rock with a hole in it to remove the thorns.

Cautions
Before you eat the petals or fruit, make sure the plants have not been sprayed with any pesticides.

Wild rose hip

BLACKBERRY/BRAMBLES
Rubus spp.

Blackberry vine in flower

There are about 400 to 750 species of *Rubus* worldwide, including at least 14 here. These include blackberry, raspberry, salmonberry, and thimbleberry (*R. parviflorus*).

Use: Berries used for juices, jams, and desserts or dried; leaves used for medicine

Range: Riparian and many other areas where sufficient water is supplied

Similarity to toxic species: Somewhat resembles poison ivy, though poison ivy lacks thorns

Best time: Fruits mature in summer

Status: Very common

Tools needed: None, but clippers can help

Properties

Even non-botanists can usually identify the vine and fruit of the very common blackberry. In some areas, wild blackberries are so common that

The blackberry vine HELEN W. NYERGES

most go uneaten. They are sometimes regarded as a nuisance.

The leaves are palmately divided (like a hand) into three, five, or seven segments. The vines twine on the ground or over low hedges and are characterized by their thorns, which make it difficult to wade too deep into any of the old hedge-like stands of wild blackberries. The flowers are white, five-petaled, and followed by the fruits, which are aggregate fruits.

Himalayan blackberry flowers—Rubus bifrons LILY JANE TSONG

Most people instantly recognize the shape of the blackberry because they've seen it in supermarkets or in back-yard gardens. The aggregate fruit is a collection of sweet drupelets, with the fruit separating from the flower stalk to form a somewhat hollow, thimble-like shape.

Uses

The blackberry is fairly universally recognized, and everyone who sees the ripe ones ventures out to eat them. I've picked them in the foothills and mountains and along roadsides. The key is to avoid the thorns and to make sure they are not immature and tart. If the fruit is black, soft, and easily picked—it's ripe! You can eat them right away or pick a bunch and mash them for a pancake, biscuit, or cake topping. Even better, sprinkle them over a bowl of vanilla ice cream. (Yes, we know that chocolate ice cream is

A view of the thimbleberry plant and fruit JEAN PAWEK

better for you, but the flavor of blackberries clashes a bit with chocolate.)

You could also make a conserve, a jam, a jelly, a pie filling, or a juice. It's a very versatile berry. And though I rarely have ripe blackberries around long enough to dry them, they can be dried in any food dehydrator and will keep for quite a while. The dried fruits can then be eaten as is or reconstituted for juices or desserts.

An infusion of the leaves has long been used among Native Americans for diarrhea and childbirth pains.

Fruiting blackberry

Salmonberry fruit VERNON SMITH

Himalayan blackberry fruit, ripe and immature LILY JANE TSONG

NIGHTSHADE FAMILY (SOLANACEAE)

There are 75 genera of the Nightshade family and 3,000 species worldwide. Many are toxic, and many are good foods.

The nightshade flowers and immature fruit

BLACK NIGHTSHADE
Solanum nigrum

There are approximately 1,500 species of *Solanum* in the world, with 13 recorded here.

Use: Fruits eaten when ripe, raw or cooked

Range: Disturbed soils and urban areas

Similarity to toxic species: According to many, this *is* a toxic species, meaning don't eat the raw green fruits, and don't eat the leaves raw. Sickness is possible in either case. There is also a slight resemblance to jimsonweed (Datura spp.), which is in the same family.

Best time: Summer

Status: Somewhat common in weedy areas

Tools needed: None

Properties

The very young plant resembles lamb's quarter; however, whereas lamb's quarter has an erect stem, nightshade is more widely branched. Also, though the individual leaves of both nightshade and lamb's quarter are quite similar, nightshade lacks the mealy coating of lamb's quarter and lacks the often noticeable red in the axil of the leaf that is common in lamb's quarter.

The five-petaled white to lavender flower is a very typical nightshade family flower, resembling the flowers of garden tomatoes. The fruits begin as tiny BB-size green fruits, and by August, they ripen into purplish-black little "tomatoes." We've eaten these fruits, always when ripe, both raw and cooked, with no problems.

A view of the nightshade fruit DR. AMADEJ TRNKOCZY

Uses

The fruit of this plant seems to peak around August when the plant can be prolifically in fruit if the season's rain and heat have been just right.

You don't want to eat these fruits raw while they are still green, as this could result in a stomachache and minor sickness. They should first be boiled, fried, or otherwise cooked. I

The ripe nightshade fruits

will, on the other hand, try a few of the dark purple ripe fruits if I see them while hiking. I like the fresh tartness. It's very much like eating a tomato but a bit spicier. They are great added to salads—just like adding tomatoes!

But just like tomatoes, there are many other ways to enjoy the ripe nightshade fruit. We've smashed them and added them to pizza dough. They taste like tomatoes but turn nearly black when cooked. They are good added to soup too. You don't need to cut or slice them since they are so small. Just toss them into your soup or stew.

Also, just like sun-dried tomatoes with their unique flavor, you can let night-shade berries dry in the sun (or in your oven or food dryer) and then eat as is or reconstitute later into various recipes. Though it isn't absolutely necessary, I find that they dry quickly if you gently smash them first—such as on the cookie sheet that you'll be drying them on.

Cautions

Some members of the Nightshade family are toxic, so do not eat any member of the family if you have not positively identified it as an edible species.

Do not eat the green berries raw. Only eat the fully ripe, dark purple berries. Otherwise, sickness could result. Green berries should only be consumed if boiled, fried, or otherwise cooked. Anyone with a tomato sensitivity, or sensitivity to other members of this family (e.g., eggplant, chilies, peppers), would do best to avoid these fruits.

Fruit and leaf of the nightshade plant HELEN WONG

NETTLE FAMILY (URTICACEAE)

The Nettle family includes 50 genera and 700 species worldwide.

A view of the nettle plant

STINGING NETTLE
Urtica dioica

Of the 45 species of *Urtica* worldwide, two are found here.

Use: Leaves used for food and for tea; stalks made into fiber

Range: Riparian, urban fields, edges of farms, disturbed soils, etc.

Similarity to toxic species: While nettle itself is regarded as a mechanical toxin by some botanists, it is safe to eat the cooked greens

Best time: Collect the greens in the spring

Status: Common

Tools needed: Gloves, snippers

Properties

This perennial generally sends up a single stalk in the late winter or spring that can grow to around five feet tall if undisturbed. The leaves are oblong, with toothed margins, and they taper to a point. Both the leaves and the stalks are covered with bristles that cause a stinging irritation when you brush against them.

Nettle can be found all over our area, along streams in the wilderness and in fields and backyards.

Uses

The young, tender leaf tips of nettle are the best to use, though you could also collect just the leaves later in the season (the stems get too tough). These tender tops can be steamed and boiled, which removes the sting of the nettles. They are tasty as a spinach-like dish, alone or served with butter or cheese or other toppings.

Also, try drinking the water from the boiling—it's delicious!

We've also made delicious stews and soups, which began by boiling the nettle tops. Then we quickly added diced potatoes, some red onions, and other greens. You can also add some miso powder. Cook until tender and then serve, perhaps with Braggs Liquid Amino Acid added for some great flavor and nutrition.

This is a vitamin-rich plant, so you'll be getting your medicine when you eat it.

Cautions

As you will probably learn from personal experience, you get "stung" when you brush up against nettle. This is due to the formic acid within each "needle," which causes skin

Nettle drying in an open-air mesh drying rack

Young nettle tops are added to soup.

Pascal's Stinging Nettle Hot Sauce

I created this hot sauce through experimentation and really enjoyed it. It has a mild "wild" flavor and was really liked by those who tasted it. It's extremely simple to make. This is a basic recipe, but you can add some of your favorite flavors and ingredients, such as Italian herbs, bay leaves, and so on. As for supplies, you'll need latex gloves, a blender (or go primitive with a knife and a molcajete), jars or bottles, and a metal pot.

5 ounces jalapeño peppers, stemmed and chopped with seeds (make sure they're not too hot, though)

1 ounce serrano peppers

5 ounces fresh nettle leaves (or young nettles)

Juice from 2 limes

6 garlic cloves

3 ½ cups apple cider vinegar

1 teaspoon kosher or pickling salt

1 cup water or white wine (I used white wine in my original recipe)

Blend all the ingredients until smooth. Strain for a thinner sauce, or keep it as is for a thicker sauce. Transfer to jars and cover. Refrigerate for at least two weeks, then enjoy!

—RECIPE FROM PASCAL BAUDAR

FORAGER NOTE: Nettles are an undervalued medicine, and herbalists speak highly of the many uses for nettle tea. I have found that drinking nettle tea in the spring helps to alleviate the symptoms of pollen allergies.

The tender leaves of the young nettle plant

irritation. So be careful when you gather nettle greens by wearing gloves or other protection. And if you do get nettle rash, you can treat it with fresh aloe vera gel or with the freshly crushed leaves of plants such as chickweed or curly dock. When I need to treat the nettle sting (usually for other people), I crush a handful of the tender chickweed plants and rub the bruised leaves over the sting. The stinging quickly diminishes.

Student Khangi examines the "needles" on the nettle plant.

Adrian Gaytan sells bundles of "hortiga" (nettles) at farmers' markets.

VIOLET FAMILY (VIOLACEAE)

There are 23 genera of this family and about 830 species, around 500 of which are *Viola*, the only genera found in this area.

A patch of Viola adunca in the wild JEAN PAWEK

VIOLET
Viola spp.

There are at least 24 species of *Viola* recorded in this area.

Use: Edible leaves and flowers

Range: Widespread, growing in most environments in our area. You can find them in fields, lawns, foothills, urban areas, etc.

Similarity to toxic species: None

Best time: Spring

Status: Common

Tools needed: Just a bag for collecting

Properties

These are commonly planted as garden plants, and they are hardy. They will spread by their roots and appear to naturalize in areas where they were once cultivated. In fact, they are very easy to cultivate if you want some nearby for your meals. We have often seen them growing in the cracks in the sidewalks in Alexandria and elsewhere.

Though there is great variety in size and minor leaf characteristics, they all have heart-shaped leaves, usually on a long stem of a few inches. The flowers are white, purple, blue, and even yellow, though the cultivated ones are purple or blue.

Wild Viola odorata JEAN PAWEK

Uses

When I learned that you could eat violets, I began by collecting the heart-shaped leaves from neighbors' yards as I walked home from school. As I came to recognize them, I noticed that they were very common on the edges of people's yards, probably just going wild from an original planting. I'd pick a few leaves here and a few there, and when I got home, I'd cook

Violets growing in the urban fringe of Alexandria

them up with a little water and season them with just butter. I loved them!

I have collected the tender leaves of spring, washed them and diced them, and added them to omelets. I've even tried some diced and added to ramen soup. They are very versatile, not strongly flavored, and can be added to many dishes.

The leaves are also edible raw, and they add their mild flavor to salads. Some people find the leaves a bit strong or tough in salads, but it's really a matter of personal preference.

The flowers are often used to make jellies, or added to jellies, as well as used in various dessert items. I have had a gelatin product that someone else made using the purple flowers, and I thought it was very tasty.

If you enjoy this tasty green, go ahead and plant some in your yard. They don't require a lot of care, and they will provide you with food year after year. We have observed one plant that seasonally grows out of a narrow crack in an asphalt driveway!

Monocots

These have one cotyledon. Leaf veins are generally parallel from the base or mid-rib, and flower parts are generally in threes.

WATER PLANTAIN FAMILY (ALISMATACEAE)

There are 12 genera worldwide of this family. (The *Alisma* genus, not treated here, is also an edible one.)

The wapato flower LOUIS-M. LANDRY

WAPATO
Sagittaria spp.

Of the 20 or so species of *Sagittaria*, at least eight have been recorded in the greater DC area. Many are easily recognized by their arrowhead leaves; others have lanceolate leaves. They always grow in permanent water.

Use: Bulbs are used, raw or cooked, but mostly cooked.

Range: Considered a North American native, these have been found throughout this area in the lower wetland elevations, along slow streams, at the edges of marshes and lakes.

Similarity to toxic species: There are some ornamental plants that bear a similarity to this arrowhead leaf, though they are not usually growing in the water.

Best time: Collect the tubers in the fall

Status: Somewhat common

Tools needed: Possibly waders or a canoe

Properties

This is a fairly easy plant to recognize with its unmistakable arrowhead-shaped leaves, always growing in swampy water or on the edges of lakes or slow streams. The little white three-petaled flowers are typically formed in whorls of three near the top of the naked stalk. The tubers are usually about the size of an egg and are white-colored. From the base of the leaves forms a network of fine fibrous roots. The tubers will develop at various distances from the base of the leaves, usually a few feet away. The tubers are white with a smooth texture.

Uses

The best time to harvest the tubers is late summer to fall, when the tubers are largest. Once you have located a patch early in the year, you can go back later and collect where you see the dried wapato stalks. There seems to be no easy way to harvest these other than wading into the mud and separating the tubers from the fine roots with your feet.

I have tried simply pulling up the whole plant, and sometimes this works too. But usually, I was only able to collect these by wading into the muddy water, loosening the tubers from the roots with my feet and toes, and then collecting them by hand.

Some of the Indian people were known to have collected the wapato tubers by going into the wet areas in a canoe and pulling the plants up from the canoe.

Collect the bigger tubers and leave the rest. The tubers can average

The arrowhead-shaped wapato leaf LOUIS-M. LANDRY

The flowering wapato plant VICKIE SHUFER

about an inch in diameter and can be as small as a marble.

Once they are washed, you use them in any of the ways you might use potatoes: boiled, fried, and/or baked. They are tasty, though the similarity to actual potatoes is slight. Though these can be eaten raw, they sometimes impart an irritation to the throat. To be safe, if you want to use the tubers in salad, boil them first, chill, and then add other salad ingredients like tomatoes, hard-boiled eggs, dressing, and so on.

Tubers from wapato KYLE CHAMBERLAIN

The wapato plant in water

ONION (OR GARLIC) FAMILY (ALLIACEAE)

There are 13 genera and 750 to 800 species of the Onion family worldwide. (Some texts still refer to this group as part of Amaryllidaceae or Liliaceae.)

Allium tricoccum, commonly called "ramps"

WILD ONIONS
Allium spp.

There are about 750 to 800 species of *Allium* worldwide. These are found widely in North America, with at least nine found in the area covered by this book. Found commonly is the small *Allium vineale* (field garlic, or crow garlic, sometimes called King Street chives) or *Allium canadense*.

Use: All tender portions eaten raw or cooked

Range: Can be found widely in lawns and fields

Similarity to toxic species: See Cautions

Best time: The leaves and flowers are most noticeable in the spring and early summer

Status: Common and abundant in its ideal habitats

Tools needed: None

Properties
Wild onions go by many names: ramps, wild garlic, leeks, etc. In general, they look like small "green onions" from the market, though many are inconspicuous when not in flower.

There is a small underground bulb, and the leaves are green and hollow. The flower stalk tends to be a bit more fibrous than the leaves. There appear to be six petals of the same color, but in fact, there are three sepals underneath the three identical petals, giving the appearance of a six-petaled flower. The expedient field key to identifying a wild onion is the unmistakable aroma. If you don't have that aroma, you shouldn't use the plant because similar-appearing members of the Lily family could be toxic or poisonous.

Tufts of wild onions HELEN W. NYERGES

Wild onions can be found all over the United States in a broad diversity of eco-types. When we finally noticed some in a Virginia lawn, we began to notice them "everywhere," hidden in plain view. We found them in the Mall at the Vietnam Veterans Memorial, and they are quite widespread.

They are noticed mostly when they flower, because otherwise, they appear very much like grass.

The common *A. vineale* originated in North Africa, Southwest Asia, and Europe, and it has been introduced into North America, where it is found over a wide area. This one has a small tuber, and the leaves are often confused for grass.

Uses

When you find wild onions, you'll be tempted to pull up the plant so you can eat the bulb. That's what you probably would do in your own garden, but that's not the only way you can use these. Generally, I only pick the green leaves for consumption. If there are a lot of them, I might take some of the bulbs to eat and break up the cluster and replant some. The reason that I generally only eat the greens is that I've seen some patches of wild onions disappear entirely due to foragers uprooting the whole plant.

So, while the wild onion bulbs can be used in any of the myriad ways in which you're used to eating garlic, onions, chives, leeks, and so on, you'll still get most of the flavor and most of the nutritional benefits by eating only the leaves. I pinch off a few leaves here, a few there, and add them to salads. Diced, they're great in soups, stews, egg dishes, and stir-fries. And, if you ever have to live off MREs, you can spice them up, and add to their nutritional value, by adding wild onion greens.

All tender parts of wild onions are edible, above and below ground. Generally, the older flower stalks become fibrous and unpalatable. Otherwise, the bulbs and leaves are all used raw or cooked. Simply remove any outer fibrous layers of the plant, rinse, and then use in any of the ways you'd use green onions or chives.

Wild onions can be added to salads, used as the base for a soup, cooked alone as a "spinach," chopped and mixed into eggs, cooked as a side to fish, and used to enhance countless other recipes. Wild onions share many of the healthful benefits of garlic and improve any urban or wilderness meal. Backpackers who are relying on dried trail rations will certainly enjoy the sustenance of wild onions. Many American Indians heavily relied on wild onions and regarded them as a staple, not just a condiment.

Excellent health benefits are associated with eating any members of this group. Some of these benefits include the lowering of cholesterol levels, prevention of flu, and reduction of high blood pressure. Used externally, the crushed green leaves can be applied directly to wounds to prevent infection.

In the New Forest in Hampshire, a woman claimed to have cured herself of tuberculosis by living on a diet of *A. vineale*, bread, spring water, and not much else. This was reported by Henry Chichester Hart in his 1898 book, *Flora of the County Donegal*.

Onions growing on a lawn blend invisibly into the grass. HELEN W. NYERGES

In Warwickshire, the dried bulbs are crushed to a powder and then worn on a flannel inside the shoe as a way to cure colds, according to Larch S. Garrad in the 1985 book *A History of Manx Gardens*.

Cautions

Never forget that some members of the Lily family with bulbs can be poisonous if eaten. Wild onions used to be classified in the Lily family because their characteristics are so similar. Make absolutely certain that you have correctly identified any wild onions that you intend to eat. You should check the floral characteristics to be certain that there are three sepals and three petals. Then, you must detect an obvious onion aroma. If there is no onion aroma, do not eat the plant. Though there are a few true onions that lack the onion aroma, it is imperative that you have absolutely identified those non-aromatic species as safe before you prepare them for food.

ASPARAGUS FAMILY (ASPARAGACEAE)

There are three genera in this relatively new family, created out of the Lily family (*Liliaceae*). Asparagaceae contains about 320 species, the majority of which (about 300) are a part of the *Asparagus* genus.

A bundle of asparagus spears

WILD ASPARAGUS
Asparagus officinalis

Of the approximately 300 species of *Asparagus*, this is the only one recorded in our area.

Use: Young shoots eaten

Range: Found in disturbed soils, valleys, grasslands

Similarity to toxic species: See Cautions

Best time: Spring

Status: Common

Tools needed: Knife, bag

Properties

The edible part of this European native is the first spring shoots, which are identical to the cultivated plant. Have you ever seen an asparagus spear in the produce store or at a farmers' market? Now you know what wild asparagus looks like! The wild asparagus can be an escapee from gardens and farms and would not be uncommon along a road or trail.

As the asparagus shoot continues to grow, numerous stems grow out of the main shoot. As these stems and their ferny leaves mature, the overall appearance of the plant begins to resemble a three-to-five-foot-tall ferny bush. As the shoots grow, they become intricately branched, giving the entire plant a ferny appearance. Eventually, the plant develops quarter-inch-long, bell-like green flowers that are followed by small berries, dark green at first, then maturing to red.

A view of the leaf and fruit of wild asparagus JIM ROBERTSON

Uses

Wild and cultivated asparagus are more or less identical. The wild shoots can be used in all the ways in which you'd use store-bought asparagus. They can be steamed or boiled and served with butter, cheese, or whatever. They can be made into soup or added to soups and stews and even eaten raw in salads.

When you've identified wild asparagus, you can go back to that patch spring after spring to collect the young shoots. A healthy asparagus patch can live for 30 to 50 years. It's a long-lived perennial. For this reason, asparagus is one of the best plants to grow in the lazy gardener's self-sustaining garden plot.

The red fruit of the mature asparagus plant DR. AMADEJ TRNKOCZY

The plant is inedible once it has grown to the point of being highly branched.

Cautions

Eating raw asparagus shoots and the small red berries causes a mild dermatitis reaction in some individuals.

Do not eat the red berries of the maturing plant. Consume only the newly emerging shoots.

GRASS FAMILY (POACEAE)

There are 650 to 900 genera worldwide, with about 10,550 species. A massive group! There are so many species that the family is divided into five or six major categories, depending on the botanist. (Some botanists are joiners, some are splitters, and the splitters seem to be getting the upper hand.) In our area, there are over a hundred genera and hundreds of individual species.

The Grass family has the "greatest economic importance of any family," according to botanist Mary Barkworth, citing wheat, rice, maize, millet, sorghum, sugarcane, forage crops, weeds, and thatching, weaving, and building materials.

A view of one of the many common wild grasses, here seeding

Use: Leaves for food (sprouts, juiced, etc.); seeds for flour or meal

Range: Grasses are truly found "everywhere"

Similarity to toxic species: See Cautions

Best time: Somewhat varies depending on what grass we're talking about, but generally spring for the greens and summer to fall for seed.

Status: Very common

Tools needed: None

Properties

Over 300 individual species of grass have been recorded in the area covered by this book.

The large plant family *Poaceae* (formerly *Gramineae*) is characterized by mostly herbaceous but sometimes woody plants with hollow and jointed stems, narrow sheathing leaves, petal-less flowers borne in spikelets, and fruit in the form of seedlike grain. It includes bamboo, sugarcane, numerous grasses, and cereal grains such as barley, corn, oats, rice, rye, and wheat.

Wild oats is another of the common grasses.

Grasses are generally herbaceous. They can be little annuals to giant bamboos. The stems are generally round and hollow, with swollen nodes. The leaves are alternate, generally narrow, linear sheathing leaves, with petal-less flowers formed in spikelets and fruit in the form of seedlike grain. The flowering and seed structures are rather diverse, ranging from the sticky seeds of the foxtail grasses that get caught in your socks to the open clusters of sorghum to such seeds as rice and wheat and the cobs of corn. Indeed, whole books have been written describing the diversity of this large family.

One common example is the common reed, or *Phragmites australis*.

Grass-like plant families that are often confused for grasses include cattails, *Juncus* species, rushes, and sedges, none of which are grasses.

Harvesting wild grass seed must be done at the right time.

Uses

The edibility of the wild grasses, generically, can be summed up in two categories: the young leaves and the seeds.

You may have had some of the leaves when you went to a health food store and ordered "wheatgrass juice." That's perhaps one of the best ways to eat various grass leaves—juice them. You can purchase an electric juicer or a hand-crank juicer. I have juiced various wild grass leaves and found the flavor to be quite diverse. Some have the flavor of wheatgrass juice and are good added to drinks or to soup broth. Some are very different, almost like seaweed, and these are typically better in soup.

However you do it, get the grasses as young as possible. They are most nutritious at this stage and are less fibrous. You will discover that grasses contain *a lot* of fiber once you start to crank a hand juicer and watch as the green liquid gold come out one end and the strands of fiber come out the other end. If you don't have a juicer, you could eat the very young grass leaves in salads or cooked soups, though you may find yourself chewing and spitting out fiber.

The seeds of all grasses are theoretically edible, though harvesting them is very difficult—if not next to impossible—in some cases. Some grass seeds are easy to collect by hand. They are then winnowed. Some are very easy to winnow of the outer chaff; some are more problematic. I have put "foxtail" grass seeds in a small metal strainer and passed them through a fire in order to burn off the outer covering. Though I was left with a little seed, I found this method less fruitful than simply locating other grasses with more readily harvestable seeds.

So even though there are 300-plus species of grass in our area, that doesn't mean that each and every one will provide you with food. The young leaves, in general, would all be a source of food, though many would have seeds that are difficult to harvest. If you are going to begin experimenting with eating grasses, you might do best to learn a few common useful grasses at first.

The seeds you gather for food should be mature and have no foreign growths on them. Then you either grind them into flour for pastry products (e.g., bread, biscuits, etc.) or cook into mush like a cereal mush.

With thousands of species worldwide on every landmass, and large numbers found here, the grasses are a group that we should get to know better. Not only are they arguably more important than trees in holding the earth together—their combined root systems are vast—but they are a valuable food source, assuming you are there at the right time to harvest the seed or leaf.

Cautions

Be aware that many substances are added to lawns and golf courses to keep the grasses green and bug-free. Those grasses *could* get you sick, so harvest with caution and common sense. Also, make sure that any seed you harvest is mature and free of any mold—mold will typically give the grain a color, such as green, white, or black. Do not eat moldy grass seeds.

CATTAIL FAMILY (TYPHACEAE)

The Cattail family contains two genera and about 32 species worldwide.

Seeding cattail spikes, though inedible in this stage, are good for staunching wounds, insulation, and tinder. RICK ADAMS

CATTAIL
Typha spp.

The *Typha* genus contains about 15 species worldwide, with three of those species recorded in this area, all very similar appearing.

Use: Food (inner rhizome, young white shoots, green female spike, yellow male pollen); leaves excellent for fiber craft where high tensile strength is not required

Range: Wetlands

Similarity to toxic species: None

Best time: Generally, the shoots and spikes are best collected in the spring. The rhizome could be collected at any time.

Status: Common in wetlands

Tools needed: Clippers, possibly a trowel

Properties

Everyone everywhere knows cattail—think of it as that grassy plant in the swamps that looks like a hot dog on a stick. Always growing in slow-moving waters or the edges of streams, these have long, flat leaves and grow up to six feet and taller. The long leaves grow from the underground horizontal rhizomes. When the plants flower in the spring, the flower spike is green, with yellowish pollen at the top. As it matures, the green spike ripens to a brown color, creating the familiar fall decoration: the hot dog on the stick.

Uses

Euell Gibbons used to refer to cattails as the "supermarket of the swamps," which is a good description of this versatile plant. There are at least four good food sources from the cattail, which I'll list in order of my preference.

In the spring, the plant sends up its green shoots. If you get to them before they get stiff and before the flower spike has started, you can tug them up, and the shoot breaks off from the rhizome. You then cut the lower foot or so and peel off the green layers. The inner white layer is eaten raw or cooked. It looks like a green onion, but the flavor is like cucumber.

The spike is the lower part of the flower spike, technically the female part of the flower. You find the spike in spring when it's entirely green and tender. Though you could eat it raw, it's far better boiled. Cook it, butter it, and eat it like corn on the cob. Guess what? It even tastes like corn on the cob. You could also scrape off the green edible portion from the woody core and add to stews or stir-fries or even shape into patties (with egg or flour added) and cook like burgers.

Green cattail spikes RICK ADAMS

David Martinez examines mature brown cattail spikes.

The pollen is the fine yellow material that you can shake out of the flower spikes. The flower spike is divided into two sections: the lower female part, which can be eaten like corn on the cob, and directly on top, the less substantial male section, which produces the fine yellow pollen. If you're in the swamp at the right time, typically April or May, you can shake lots of pollen into a bag and then strain it (to remove twigs and bugs) and use it in any pastry product.

The white edible shoots at the base of the cattail stalk

The rhizome is also good starchy food. Get into the mud and pull out the long horizontal roots. Wash them, and then peel off the soft outer layer. You could just chew on the inner part of the rhizome if you need the energy from the natural sugar, or you could process it a bit. One method of processing involves mashing or grinding up the inner rhizome and then putting it into a jar of water. As the water settles, the pure starch will be on the bottom, and the fiber will be floating on the top, so you can easily scoop it out and discard it. The starch is then used in soups or in pastry and bread products.

The white shoots are being prepared to be eaten raw in salads and in various cooked dishes.

Aside from cooking, the long green leaves have a long history of being used for various woven products that will not be under tension, such as baskets, sandals, and hats, and even for the outer layers of the dwellings utilized by many of the West Coast Native Americans.

And when that cattail spike matures to a chocolate-brown color, it can be broken open, and it all turns to

Angelo Cervera shows some ready-to-cook green cattail spikes.

an insulating fluff. Each tiny seed is actually connected to a bit of fluff that aids in the transportation of that seed to greener grass on the other side. You can use that fluff to stop the bleeding of a minor wound, as an alternative to down when stuffing a sleeping bag or coat, and as a fantastic fire starter!

OTHER EDIBLES

Do you have a favorite wild food or wild plant that wasn't listed here? Sedges, buckeye, various Mustard family greens, various nuts, elm seeds, Osage orange seeds, tuckahoe, evening primrose, mullein, willow, poison ivy, and so on?

Remember, our intent was to include those *food* plants that are the most widespread throughout the area and readily recognizable and those that would make a significant contribution to your day-to-day meals. We didn't include any endangered or rare species. This book was compiled based on what we ascertained were the plants that you are most likely to be eating from the wild, most of the time.

Malcolm McNeil shared with us the book *The Powhatan Indians of Virginia* by Helen C. Rountree, which quotes a historical source about the early diet: "In March and April they live much upon their [fishing] Weeres, and feed on Fish, Turkeys, and Squirrells and then as also sometymes in May they plant their Feilds and sett their Corne, and live after those Monethes most[ly] of[f] Acrons [sic], Wallnutts, Chesnutts, Chechinquamyns and Fish, but to mend their dyett, some disperse themselues in smale Companies and live vpon such beasts as they can kill, with their bowes and arrows. Vpon Crabbs, Oysters, Land Tortoyses, Strawberries, Mulberries, and such like; in Iune, and Iuly, and August they feed vpon the rootes of Tockohowberryes [wild potatoes], Grownd-nuts, Fish, and greene Wheat [corn], and sometime vpon a kynd of Serpent, or great snake of which our people likewise vse to eate."

A lot of the foods eaten by early peoples of this area are easily recognized, such as onions, carrots, grapes, acorns, walnuts, strawberries, and mulberries, and most are in this book.

Then there is "tuckahoe," which led us on a wild research journey. The term Tuckahoe can refer to the Indigenous people of the Virginia area or in the area of Tuckahoe, New York. The word can refer to fungus and at least two members of the *Araceae* family: either *Orontium aquaticum*, commonly called "floating arum," or *Peltandra virginica*, commonly called "arrow arum." Apparently, the local Indigenous people ate one or both of these roots. I don't spend time describing them for everyday foragers because, as members of the *Araceae* family, they contain calcium oxalate crystals and require long-term cooking, drying, or heating in order to render the crystals harmless. I don't doubt that people would eat these when they had no other foods, but they are clearly a *long-term survival food* and not the sort of food that you'd bother with unless you really had very few other options.

As you continue your study of ethnobotany, you will discover that there are many more wild plants that could be used for food. Some are marginal, and some just aren't that great. Some might be really good but were not included because of the various limitations of putting together a book.

Yes, there are also many wild animals and ocean life that could be used for food—fish, snakes, lizards, birds, small mammals, insects, and so on—but this book is about the plants.

GETTING STARTED: EXPLORING THE FASCINATING WORLD OF WILD PLANTS

During the many field trips and classes that I have conducted since 1974, I have often been asked how I got interested in the subject of edible wild plants. More importantly, someone will want to know how they should go about learning to identify and use wild foods in the safest and quickest way possible.

Though I had very little prior knowledge of ethnobotany when my interest began (about age 12 or 13), I began to seek out local botanists from whom I could learn. I also took every class on the subject and other related subjects that I could, both in high school and college. In addition, I spent a lot of time in the fields, foothills, mountains, deserts, and beaches looking at plants and collecting little samples to take back to my growing body of mentors.

All this took time, and I learned the plants one by one, by the primary method that I still recommend: Show the plant to an expert for identification or go into the field with an expert so the plant can be identified. Then, once you've identified the plant, you can do all your research in books such as this one, and—assuming it's an edible plant—you can begin to experiment with all the ways you can eat it.

I learned some of the very common widespread plants first, and I would carefully clip samples, take them home, and try them in salad or as cooked greens. Some of the very first plants I began to eat were mustard, purslane, and watercress, as these were very common and easy to collect. Once I learned the identity of another edible plant, I would try it in various recipes over the course of the next several weeks, until I felt I "knew" that plant well, and my interests moved on to learning a new plant.

There was no quickie "rule of thumb" for knowing what I could or could not eat. There were no lazy-man rules of looking for red in the plant, the color of the berries, or whether the plant left a bad taste in my mouth. There was simply the effort to discover, to learn one new plant at a time, to utilize that plant in the kitchen, and to watch that plant throughout the growing season, so I got to know what it looked like as a sprout, growing up, maturing, flowering, going to seed, and dying. Through observation, I would be able to recognize these common floral beings even if I was driving by in a car at high speed.

So that is what you should do: seek out a mentor or mentors, study with them, take plants to them, get to know the plants, and continue forever with your learning. Then, use your books, videos, and internet references as a backup to your firsthand interaction with the plant.

Where I grew up, I lived close to the local mountain range, so hiking in the hills was my after-school or weekend choice of recreation. I began to backpack and carry a heavy load and found it unpleasant—one of those things that you just had to put up with if you wanted to backpack. But then I met a man who talked about how he learned about foods that the Indians of Northern California used in the old days. He mentioned a few specific plants, and something clicked in my brain, and I knew that was a skill I had to learn. When I returned from my backpacking trip, I began to research ethnobotany at the local library and museums, and I sought out teachers and mentors.

The fact that I was always interested in the ways of Native Americans, and in practical survival skills, helped immensely. Plus, my mother often told us of the hard times she experienced growing up on the family farm in Ohio. I knew that knowledge of wild foods was an important skill that too many of us had lost.

By January of 1974, I began to lead wild-food outings that were organized by WTI, a nonprofit organization focused entirely on education in all aspects of survival. I led half-day walks where we'd go into a small area, identify and collect plants, and make a salad and maybe soup and tea on the spot. I became an active member of the local Mycological Association and made rapid progress in learning about how to identify and use mushrooms. And I continued to take specialized classes and field trips in botany, biology, taxonomy, and ethnobotany. I spent many hours in the classroom and lab of Dr. Leonid Enari, who was the chief botanist at the Los Angeles County Arboretum in Arcadia. He was a walking encyclopedia, and he also took the time to mentor me, to answer all my plant questions, and to help me with sections of my first wild-food book.

Today, there are many more learning avenues than were available for me, such as the Internet. Most of what I learned, I learned the hard way, in spite of the fact that I had many teachers over the years. So, when I teach, I attempt to provide methods for my students to save time and learn more rapidly. In short, I try to provide an ideal learning environment that I wished I'd had.

It is to your advantage to completely disregard any of the "rules of thumb" you've ever been taught about plant identification—you know, the shortcuts for determining whether or not a plant is edible, such as: If a plant has a milky sap, it is not edible. If a plant causes irritation in the mouth when you eat a little, it is not safe to eat. If the animals eat the plants or berries, they are safe to eat. If the berries are white, they are poisonous. If the berries are black or blue, they are safe to eat. And on and on. Disregard all these shortcuts since—although often based on some fact—they all have exceptions. As Spock would say, "Insufficient data." There are no shortcuts to what is necessary: you must study, and you will need field experience.

If there is any sort of "shortcut" to the study of plants, it is to learn to recognize plant families and learn to know which families are entirely safe for consumption. Beyond that, you must learn plants one by one for absolute safety.

I strongly suggest that you take at least a college course in botany (preferably taxonomy), so you get to know how botanists designate plant families. This will enable you to look at my list of safe families and then use the books written by botanists of the flora of your region so you can check to see which plants in your area belong to any of the completely safe families. After a while, this will come easy. Eventually, you'll look at a plant, examine it, and know which family it likely belongs to.

Get a botanical flora book written for your area and study it.

Do your own fieldwork, ideally with someone who already knows the plants. Gradually, eventually, you will be using more and more wild plants for food and medicine and perhaps for soap, fiber, fire, and so on. There will be no such thing as a "weed." You will cringe whenever you see the television commercials in the spring for such noxious products as Roundup that promise to kill every dandelion on your property.

When you discover that we have ruined the Earth in the name of "modern agriculture," which produces inferior "food," you will understand the meaning of the phrase, "We have met the enemy, and he is us."

TEST YOUR KNOWLEDGE OF PLANTS

Here is a simple test that I use in my classes. Take the test for plants and mushrooms and see how you do.

1. ☐ True. ☐ False. Berries that glisten are poisonous.

2. ☐ True. ☐ False. White berries are all poisonous.

3. ☐ True. ☐ False. All blue and black berries are edible.

4. ☐ True. ☐ False. If uncertain about the edibility of berries, watch to see if the animals eat them. If the animals eat the berries, the berries are good for human consumption.

5.   Would you adhere to the following advice? State yes or no, and give a reason.

According to *Food in the Wilderness* authors George Martin and Robert Scott, "If you do not recognize a food as edible, chew a mouthful and keep it in the mouth. If it is very sharp, bitter, or distasteful, do not swallow it. If it tastes good, swallow only a little of the juice. Wait for about eight hours. If you have suffered no nausea, stomach or intestinal pains, repeat the same experiment swallowing a little more of the juice. Again, wait for eight hours. If there are no harmful results, it probably is safe for you to eat. (This test does not apply to mushrooms.)"

6. ☐ True. ☐ False. "A great number of wilderness plants are edible but generally they have very little food value." [Martin and Scott, ibid.]

7. ☐ True. ☐ False. Bitter plants are poisonous.

8. ☐ True. ☐ False. Plants that exude a milky sap when cut are all poisonous.

9. ☐ True. ☐ False. Plants that cause stinging or irritation on the skin are all unsafe for consumption.

10. The illustration to the right is the typical flower formation for all members of the Mustard family. Write out the formula:

___ petal(s); ___ sepal(s);

___ stamen(s); ___ pistil(s).

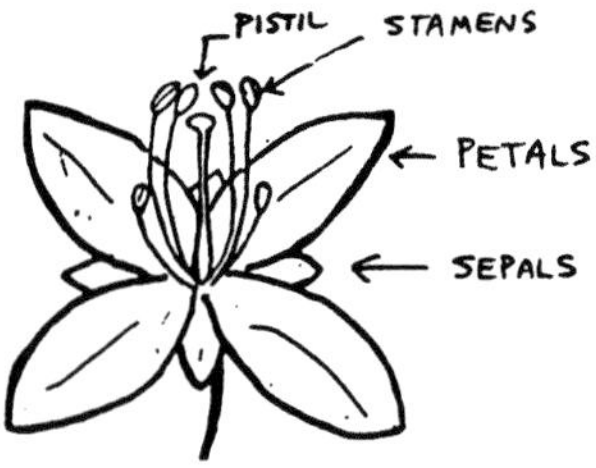

11. Of what value is it to be able to identify the Mustard family?

12. ❐ True. ❐ False. Mustard (used on hot dogs) is made by grinding up the yellow flowers of the mustard plant.

13. ❐ True. ❐ False. Yucca, century plant, and prickly pear are all members of the Cactus family.

14. ❐ True. ❐ False. There are no poisonous cacti.

15. ❐ True. ❐ False. Plants that resemble parsley, carrots, and fennel are all in the Carrot (or Parsley) family and are thus all safe to eat.

16. ❐ True. ❐ False. Only 17 species of acorns are edible. The rest are toxic.

17. To consume acorns, the tannic acid must first be removed. Why?

18. If you are eating no meat or dairy products (during a survival situation, for example), how is it possible to get complete protein from plants alone?

19. ❐ True. ❐ False. There are no toxic grasses.

20. ❐ True. ❐ False. Seaweeds are unsafe survival foods.

21. ❐ True. ❐ False. All plants that have the appearance of a green onion and have the typical onion aroma can be safely eaten.

22. List all of the plant families (or groups) from this lesson that we've identified as entirely or primarily nontoxic:

ANSWERS

1. False. Insufficient data.

2. False. Though mostly true, there are exceptions, such as white strawberry, white mulberry, and others. Don't eat any berry unless you know its identity and you know it to be edible.

3. False. Mostly true, but there are some exceptions. Don't eat any berry unless you've identified it as an edible berry.

4. False, for several reasons. Certain animals are able to consume plants that would cause sickness or death in a human. Also, animals do occasionally die from eating poisonous plants—especially during times of drought. Also, just because you watched the animal eat a plant doesn't mean the animal didn't get sick later!

5. Very bad advice, even though this has been repeated endlessly in "survival manuals" and magazine articles. Since food is rarely your top "survival priority," this is potentially dangerous advice.

6. False. To verify that this is not so, look at Composition of Foods, which is published by the US Department of Agriculture. In many cases, wild foods are far more nutritious than common domesticated foods.

7. False. Insufficient data. Many bitter plants are rendered edible and palatable simply by cooking or boiling.

8. False. Though you can't eat any of the euphorbias, many others (like dandelion, lettuce, milkweed, and sow thistle) exude a milky sap. Forget about such "shortcuts." Get to know the individual plants.

9. False. Many edible plants have stickers or thorns that must first be removed or cooked away, such as nettles, cacti, and so on.

10. Mustard flowers are composed of

    Four sepals (one under each petal)

    Four petals (the colorful part of the flower)

    One pistil (in the very center of the flower, the female part of the flower)

    Six stamens; four are tall, and two are short. (The six stamens surround the pistil.)

11. There are no poisonous members of the Mustard family.

12. False. The mustard condiment is made by grinding the seeds. Yellow is typically from food coloring.

13. False. Only the prickly pear is a cactus.

14. True, but you must know what is, and is not, a cactus. There are some very bitter narcotic cacti that you would not eat due to unpalatability. Also, some euphorbias closely resemble cacti and will cause sickness if eaten. Euphorbias exude a milky sap when cut; cacti do not. Any fleshy, palatable part of true cacti can be eaten.

15. False. The Carrot (or Parsley) family contains both good foods and deadly poisons. Never eat any wild plant resembling parsley unless you have identified that specific plant as an edible species.

16. False. All acorns can be consumed once shelled and leached of their tannic acid.

17. Tannic acid is bitter.

18. Combine the seeds from grasses with the seeds from legumes. This generally produces a complete protein. For details, see Diet for a Small Planet by Frances Moore Lappé.

Traditional Diets That Combine Legumes and Grass Seeds to Make a Complete Protein

Loosely based upon "Summary of Complementary Protein Relationships," Chart X in *Diet for a Small Planet* by Frances Moore Lappé

	Legumes	Grasses
Asian diet	Soy (miso, tofu, etc.)	Rice
Mexican diet	Beans (black beans, etc.)	Corn (tortillas)
Middle East diet	Garbanzos	Wheat
Southern United States	Black-eyed peas	Grits
Starving student	Peanut butter	Wheat bread
Others to consider	Mesquite, palo verde, peas, carob, etc.	Millet, rye, oats, various wild grasses, etc.

19. True. However, be certain that the seeds are mature and have no mold-like growth on them.

20. False. Seaweeds are excellent. Make certain they've not been rotting on the beach, and don't collect near any sewage treatment facilities.

21. True. But be sure you have an onion!

22. All members of the Mustard family, all palatable cacti, all acorns, all cattails, grasses, seaweeds, onions. There are many other "safe" families, but you will need to do a bit of botanical study in order to identify those families. Begin by reading the descriptions of each family in this book. Also, study *Botany in a Day* by Tom Elpel.

THE DOZEN EASIEST TO RECOGNIZE, MOST WIDESPREAD, MOST VERSATILE WILD FOODS OF MARYLAND, VIRGINIA, AND THE DC AREA

Since the early days of our botanical studies, we found that there are quite a few entire families that are safe to eat, given a few considerations in each case. Some of these families are difficult to recognize unless you are a trained botanist. Some of those entirely safe families are described in this book.

Christopher's original research on this was done with Dr. Leonid Enari, one of his teachers and the chief botanist at the Los Angeles County Arboretum in Arcadia, California.

The chart below was the idea of his friend Jay Watkins, who long urged Christopher to produce a simple handout on the dozen most common edible plants that everyone should know. Granted, there are many more than a dozen, but as Jay and Christopher discussed the idea, the decision was made to focus on 12 plants that could be found not just anywhere in the United States but in most locales throughout the world. The result was the accompanying chart, which is largely self-explanatory.

This chart assumes that you already know these plants since its purpose is not identification. Anyone who has studied wild foods for a few years is probably already familiar with all these plants. However, not everyone is aware that these plants are found worldwide.

This overview should help both beginners as well as specialists. It is merely a simple comparative chart, which could be expanded to many, many pages. It is deliberately kept short and simple.

	Description	Parts Used	Food Uses	Preparation	Benefits	Where Found	When Found
Acorns	Trees, with nuts set in a scaly cap	Acorns	Made into flour for bread, cake, etc.	Acorns must be leached first of bitterness	Tasty and nourishing. See Oak section in this book	Widespread	Acorns mature in the fall
Cactus	Succulent desert plants of various shapes	Tender parts; fruit	Salad; cooked vegetable; omelet; dessert; drinks	1. Carefully remove spines; 2. dice or slice as needed	Pads said to be good for diabetics; fruits rich in sugar	Dry environments	Young green pads in spring and summer; fruit in summer and fall
Cattail	Reed-like plants; fruit looks like hot dog on stick	Pollen; green flower spike; tender shoots; rhizome	Flour; cooked vegetable; salads; flour	Shake out pollen; boil; remove outer green fibrous parts; remove outer parts, crush	Widespread; versatile	Wet areas, e.g., roadside ditches, marshes	Spring through fall
Chickweed	Weak-stemmed, opposite leaves, five-petaled flower	Entire tender plant	Salads; tea	Clip, rinse, and add dressing, or make infusion	Good diuretic	Common and widespread when moisture is present	Spring and summer
Dandelion	Low plant, toothed leaves, conspicuous yellow flower	Roots; leaves	Cooked vegetable, coffee-like beverage; salads	Clean and cook; or dry, roast, grind; clean and make desired dish	Richest source of beta carotene; very high in vitamin A	Common in lawns and fields	Best harvested in spring
Dock	Long leaves with wavy margins	Leaves; stems; seeds	Salads, cooked vegetable; pie; flour	Clean; use like rhubarb; winnow seeds	Richer in vitamin C than oranges	Common in fields and near water	Spring through fall

Grasses	Many widespread varieties	Seeds; leaves	Flour, mush; salads; juiced; cooked vegetable	Harvest and winnow; harvest, clean, and chop	Easy to store; rich in many nutrients	Common in all environments	Fall and spring
Lamb's Quarter	Triangular leaves with toothed margins, mealy surface	Leaves and tender stems; seeds	Salads; soups; omelets; cooked; bread; mush	Harvest and clean; winnow	Rich in vitamin A and calcium	Likes disturbed rich soils	Spring through fall
Mustard	Variable leaves with large terminal lobes; four-petaled flowers	Leaves; seeds; some roots	Salads; cooked dishes; seasoning	Gather; clean; cut as needed	Said to help prevent cancer	Common in fields and many environments	Spring through fall
Onions	Grass-like appearance; flowers with three petals, three sepals	Leaves; bulbs	Seasoning; salads; soups; vegetable dishes	Clean and remove tough outer leaves	Good for reducing high blood pressure and high cholesterol level	Some varieties found in all environments	Spring
Purslane	Low-growing succulent; paddle-shaped leaves	All tender portions	Salads; sautéed; pickled; soups; vegetable dishes	Rinse off any soil	Richest source of omega-3 fatty acids	Common in parks, gardens, disturbed soils	Summer
Seaweeds	Marine algae	Entire plant	Flavoring agent, thickening agent, for soups, etc.	Clean, and prepare fresh or dried, depending on species	Rich source of many minerals	Collected in the ocean and on the beach	Year-round

Latin names: Acorns = Quercus spp.; Cattail = Typha spp.; Chickweed = Stellaria media; Dandelion = Taraxacum officinale; Dock = Rumex crispus; Grasses = Poaceae (Grass family); Lamb's quarter = Chenopodium album; Mustard = Brassica spp. / Mustard family = Cruciferae; Onions = Allium spp.; Purslane = Portulaca oleracea

STAFF OF LIFE:
BEST WILD-FOOD BREAD SOURCES

Baking of bread goes back to the most ancient cultures on the Earth, back when mankind discovered that you could grind up the seeds of grasses, add a few other ingredients, let it rise, and bake it. There are countless variations, of course, but bread was once so nutritious that it was called the "staff of life."

Most likely, the discovery of bread predated agriculture since the Earth was full of wild grasses and a broad assortment of roots and seeds that could be baked into nutritious loaves. Most grains store well for a long time, which allowed people the time to pursue culture, inner growth, and technology. The development of civilizations and the development of agriculture go hand in hand. And bread was there right from the beginning.

Today, we are at another extreme of a very long road of human development. We started with the struggle for survival and with the surplus of the land allowing us the time to develop more fully in all aspects. That good bread from the Earth was heavy, rich, and extremely nutritious. It was a vitamin and mineral tablet.

We produced so much grain that the United States called itself "the bread-basket of the world." And this massive volume resulted in losses in the fields from insects and loss due to spoilage. Thus came the so-called "Green Revolution," where chemical fertilizers replaced time-honored ones such as animal manures, straw and hay, compost, bone meal, and other such natural substances that the modern farmer was too busy and too modern to use. Crops increased while the nutritional values dropped. And though this is a gross oversimplification, bread from the supermarket is no longer the staff of life.

The mineral content of the once-rich soils of the United States has steadily declined. Producers process and refine "white flour" and then add certain minerals back into the bread dough. We sacrificed quality since we thought it would bring us security, and we knew it would bring big bucks. Now, the great irony is that we lost the quality of the food, of the soil, and, ultimately, we are no more secure than ever before. Why? Because soil rich in natural organic matter can withstand floods and droughts and the ravages of insects. It is the folly of man who causes the droughts and plagues of modern times.

There is much—very much—that we need to learn about "modern agriculture," or "agribiz," as it is more appropriately called. We should not put our heads in the sand, ostrich-like, and pretend the problem does not exist.

Personal solutions are many. Grow your own garden. Learn about wild foods and use them daily. By using common wild plants, you can actually create a nutritious bread comparable to the breads your ancestors ate. The easiest way to get started is to make flour from these wild seeds and mix that flour fifty-fifty with your conventional flours, such as wheat. You'll end up with a more flavorful, more nutritious bread, pancake, or pastry product.

Once you begin to use your local wild grains, you'll be amazed at how tasty, how abundant, and how versatile these wild foods are.

The accompanying chart is by no means complete. It is a general guideline to show you what is available over widespread areas. However, there are quite a few plants of a limited range that produce abundant seeds or other parts that are suitable for bread making. In most cases, you should consult any of the many wild-food cookbooks available for details on using each of these wild grains.

Note that "Grass" is a huge category since it actually includes many of our domestic grains such as wheat, corn, rye, barley, and so on. Though some of the seeds listed in this chart can be eaten raw, most require some processing before you can eat them. The seed from amaranth, dock, and lamb's quarter can get somewhat bitter and astringent as it gets older and is improved by cooking.

By rediscovering the wealth of wild plants that are found throughout this country, we can bring bread back to its status as the "staff of life."

RECIPE

Beginner Wild Bread Recipe

1 cup whole wheat flour

1 cup wild flour of your choice

3 teaspoons baking powder

3 tablespoons honey

1 egg

1 cup milk

3 tablespoons oil

Salt to taste, if desired

Mix all the ingredients well and bake in oiled bread pans for about 45 minutes at 250°F or in your solar oven until done.

Beginner Pancake Recipe

Follow the above recipe, adding extra milk or water so you have pancake batter consistency. Make pancakes as normal.

"Wild Bread" Chart

	Part Used	How Processed	Where Found	Palatability	Ability to Store
Acorns	Shelled acorns	Leach acorns of tannic acid by soaking or boiling, and grind into meal	Worldwide; ripens in fall	Good, if fully leached	Excellent
Amaranth	Seeds	Collect and winnow seeds	Worldwide as a weed of disturbed soils	Good	Very good
Cattail	Pollen and rhizome	Shake the top of cattail spikes into bag to collect pollen; mash peeled rhizome and separate out fiber	Worldwide in wet and marshy areas	Very good	Good
Dock	Seeds	Collect brown seeds in fall, rub to remove "wings," and winnow	Worldwide in wet areas and disturbed soils	Acceptable	Very good
Grass— most species	Seeds	Generally, simply collect and winnow; difficulty depends on species	Worldwide; some found in nearly every environment	Generally very good	Very good to excellent
Lamb's Quarter	Seeds	Collect when leaves on plant are dry; rub between hands and winnow	Worldwide in disturbed soils and farm soils	Acceptable to good	Very good

Note: This chart is intended only as a general guideline to compare sources for "wild bread" ingredients. There may be many other wild plants that can be used for bread. Also, never eat any wild plant that you have not positively identified as an edible species.

Latin names: Acorns = Quercus spp.; Amaranth = Amaranthus spp.; Dock = Rumex crispus; Grass = Poaceae; Lamb's quarter = Chenopodium album

SWEET TOOTH:
BEST WILD-FOOD SUGARS AND DESSERTS

When people speak of "sugar" today, they are almost always talking about the highly refined, nutritionless, white substance made from sugarcane or sugar beets. Unfortunately, modern sugar is a foodless food. It is the "cocaine" of modern man's dinner plate. It is not good for the body, and it offers no nutrients whatsoever. But this has not always been so.

Just a few generations ago, it was common for people to make their own sugars. Every culture had its favorite sources for its sugars, depending on what was found in the wild or what was grown in that particular location. In most cases, they simply collected, dried, and ground up sugar-rich fruits. Most such fruits naturally crystallized with time and then could be further ground. The advantage of these sugars over "white cane sugar" is that these sugars had their own individual flavors, and they contained many valuable minerals.

Some sugars are quite simple to "produce," such as honey. The main obstacles are to find a way to house the bees—something modern beekeepers do quite well—and preventing getting stung. And tapping maple trees (and several other trees) was so simple that even the North American Indigenous peoples did it with relatively primitive tools. They simply cut narrow slashes into the tree, inserted hollow tubes made from elder branches, and collected the sap in whatever containers they had. Raw maple sap is usually boiled down to get a syrup of desired consistency and sugar content. Sometimes, you boil off about 40 gallons of water for each gallon of syrup. You do *not* do this indoors.

People have always sought ways to make foods more flavorful, and sugar is certainly useful in that regard. But sugar is also valuable as a preservative. Both sugar and salt help to preserve foods and keep them from spoiling. This was especially important in the past when there was no electricity or refrigerators.

It's amazing how fast a modern culture forgets things. Probably not one in a thousand urbanites knows these simple details about sugar. Our culture has sunk to such ignorance in this matter that we somehow believe that the only choice is between white sugar, the pink container, or the blue container. Rather than produce nutritious sugars as we had in the past, the trend is to produce high-tech sweet substances that not only have no nutrients but have no calories either, as does white sugar. The wonders of science never cease!

For those of you who want to try making sugar, the accompanying chart gives you some ideas as to what is available. Currants and gooseberries were very popular among Indigenous peoples not as a sweetener but as a preservative. They ground up their jerky and added crushed currants or gooseberries, and the result was pemmican.

Though many of the wild berries described in the chart have been used as sweeteners for other foods, most of them are good foods in their own right and have long been used to make such things as drinks, pies, jams, custards, and a variety of dessert items. Details for these can be found in many of the wild-food cookbooks available.

Here's one recipe, which can be used with all of the sugars in the chart.

RECIPE

Wild Brickle

½ gallon ripe blackberries

Water to cover berries

½ cup honey

Approximately ⅓ cup biscuit mix

Begin by gently cooking the washed berries. When they are cooked, add the honey and stir. After the mixture thickens, stir in the biscuit mix little by little. The mix will be very thick when it is ready to serve.

This makes a heavy, sweet dessert. In the old days, what wasn't eaten would be put into a bread pan and baked until dry so it would store. It would then last a long time until reconstituted. The dried shape looks like a brick, which is the source of the name.

"Wild Sugar" Chart

	Part Used	How Processed	Where Found	Sweetness/ Palatability	Ability to Store
Apples, Wild (including crabapples)	Whole fruit	Use fresh, or slice, dry, and grind to flour	Entire US	Very good; collect when ripe	Very good
Berries (blackberries, raspberries, thimbleberries)	Whole fruit	Use fresh, or dry and grind	Entire US	Excellent	Very good
Currants	Whole fruit	Use fresh, or dry and store	Entire US	Good	Very good
Elder	Whole fruit	Can use fresh if cooked first, or dry and store	Entire US	Contains sugar but tart	Good
Gooseberries	Whole fruit	Remove spiny layer, then use fresh or dry	Entire US	Good	Very good
Grapes, Wild	Whole fruit	Use fresh, or dry	Most of the US	Sometimes tart; collect ripe fruit	Very good
Maple	Sap	Cut bark on tree and capture sap; use fresh; crystallizes naturally	Entire US, but best flow where there is snow	Excellent	Excellent
Prickly Pear Cactus	Fruit	Remove stickers, use inner pulp fresh or dried, with or without seed	Entire US, but most common in Southwest	Excellent	Good

Note: Many sugars are found in nature, usually in the fruits. Honey is a traditional sugar, made indirectly from plant nectars. Other traditional sugars include dried and powdered dates, dried pomegranate juice, and beets. This chart compares a few wild sugar sources that are the most widespread throughout North America. There are many plants that are either marginal sugar sources or available in very limited locations. Never use any wild plant for sugar or food until you have positively identified it as an edible plant.

Latin names: Apples, wild = Malus spp.; Berries = Rubus spp.; Elder = Sambucus spp.; Gooseberries = Ribes spp.; Grapes, wild = Vitis spp.; Maple = Acer spp.; Prickly pear cactus = Opuntia spp.

USEFUL REFERENCES

Online References

Annotated Checklist of the Vascular Plants of the Washington-Baltimore Area, Part 1, Ferns, Gymnosperms, Dicots. Naturalhistory.si.edu/sites/default/files/media/file/chklst1.pdf.

Annotated Checklist of the Vascular Plants of the Washington-Baltimore Area, Part 2, Monocots. Naturalhistory.si.edu/sites/default/files/media/file/checklistpt2a-2a.pdf.

Plant Native. Native Plant List, Maryland, Virginia, West Virginia. www.plant-native.org/rpl-mdvawv.htm.

Maryland Biodiversity Project. www.MarylandBiodiversityProject.com.

Books

Angier, Bradford. *Free for the Eating.* Mechanicsburg, PA: Stackpole Books, 1996.

Baldwin, Bruce G., et al., eds. *The Jepson Manual, Vascular Plants of California.* 2nd ed. Berkeley, CA: University of California Press, 2012. This is the book botanists of California use, and a vast majority of the plants in Idaho can also be found in California.

Elpel, Tom. *Botany in a Day: The Patterns Method of Plant Identification.* Pony, MT: HOPS Press, 2000. Highly recommended. This is the way that botany should be taught.

Enari, Dr. Leonid. *Plants of the Pacific Northwest.* Portland, OR: Binfords and Mort, 1956. Written by Dr. Enari after he moved to Portland from Estonia, this book covers 663 weeds, wildflowers, shrubs, and trees that are common in the Northwest. Includes 185 line drawings, which means that most plants are not illustrated.

Kimmerer, Robin Wall. *Braiding Sweetgrass: Indigenous Wisdom, Scientific Knowledge, and the Teaching of Plants.* Minneapolis, MN: Milkweed Editions, 2015. Kimmerer gives you a unique perspective as an Indigenous person who's also a trained botanist. The book opens your eyes to the what, the how, and the why.

Krebil, Mike. *The Scouts Guide to Wild Edibles: Learn How to Forage, Prepare, and Eat 40 Wild Foods.* Pittsburgh, PA: St. Lynn's Press, 2016. An excellent and easy-to-follow guide to some of our most common edibles. Lots of color photos and recipes.

Krebil, Mike. *A Forager's Life: Reflections on Mother Nature and My 70+ Years of Digging, Picking, Gathering, Fixing and Feasting on Wild Edibles.* Pittsburgh, PA: St. Lynn's Press, 2021. A delightful read that documents Krebil's life as a forager, along with color photos and more recipes.

Moerman, Daniel E. *Native American Ethnobotany.* Portland, OR: Timber Press, 1998. Nearly a thousand pages of descriptions of how every plant known to be used by Native Americans was utilized. No pictures at all, but an incredible resource all in one book.

Nyerges, Christopher, *Foraging Wild Edible Plants of North America.* Guilford, CT: Falcon Guide. This is Christopher's book that includes useful plants that are found throughout the country. Full-color photos with plenty of recipes.

Rountree, Helen C., *The Powhatan Indians of Virginia.* Norman, OK: University of Oklahoma Press, 1989.

INDEX

ABOUT THE AUTHOR

Christopher Nyerges, a cofounder of the School of Self-Reliance, has led wild-food and survival skills walks for thousands of students since 1974. He has authored 22 books, mostly on wild foods, survival, and self-reliance, and thousands of newspaper and magazine articles. He continues to teach where he lives in Los Angeles County, California. More information about his classes and seminars is available at www.SchoolofSelf-Reliance.com or by writing to School of Self-Reliance, Box 41834, Eagle Rock, CA 90041.